Industrial Statistics with Minitab

Industrial Statistics with Minitab

Pere Grima Cintas
Lluís Marco Almagro
Xavier Tort-Martorell Llabrés

Universitat Politècnica de Catalunya - BarcelonaTech
Barcelona, Spain

A John Wiley & Sons, Ltd., Publication

Library of Congress Cataloging-in-Publication Data

Grima Cintas, Pere.
   Industrial statistics with Minitab / Pere Grima Cintas, Lluis Marco-Almagro, Xavier Tort-Martorell
Llabres.
     p. cm.
   Includes index.
   ISBN: 978-0-470-97275-5 (cloth)
  1. Industrial statistics.   2. Industrial management–Statistical methods–Computer programs.
3. Statistics–Computer programs.   4. Minitab.   I. Marco-Almagro, Lluis.   II. Tort-Martorell Llabres,
Xavier.   III. Title.
   HB137.G72 2012
   005.5′5–dc23   2012016129

A catalogue record for this book is available from the British Library.

ISBN: 978-0-470-97275-5

Typeset in 10/12pt Times by Aptara Inc., New Delhi, India
Printed and bound in Singapore by Markono Print Media Pte Ltd

Contents

Preface

This book is aimed at students and professionals wishing to use Minitab® Statistical Software as a tool for performing statistical analysis. The content is full of examples and applications around quality control and improvement situations, but we believe that it can be equally useful to people working in other areas who need to analyze data, especially in industrial environments.

Let us start by saying that Minitab includes a very good help facility that allows an easy and quick topic location and that they are, almost always, presented in a clear and concrete way and with good examples. Therefore, the aim of the book is different: it is to provide guidance in the use of Minitab for solving statistical problems as well as in solving problems using statistics and Minitab. Our contribution lies in the selection of materials, the structure and order in which they are presented, and a very visual way of presenting them that facilitates understanding the way to do things without reading long paragraphs. In addition the selection of examples and case studies cover a wide range of common industrial situations.

The book is divided into six parts corresponding to six groups of more or less homogeneous topics and three appendices. Each part consists of several chapters explaining how to solve particular situations – how to use a particular statistical technique – using Minitab, and closes with one last chapter dedicated to case studies. We have kept the chapters on how to use Minitab short and specific. Their content is reflected in the chapter titles, and they go straight to the point. All procedures are explained through examples, and a good way to learn is to try to reproduce them. Hints and tips to facilitate tasks or attention calls to avoid errors are highlighted and identified using icons. In the case study chapters the protagonist is the problem they present and Minitab is just the tool that helps to solve them. Naturally, not all options or techniques are discussed, but only those we consider most useful, commonly used or that serve to give an overview of more complex issues. As said before, Minitab has a first-rate help facility incorporated so that readers can explore other options by themselves. Once the reader is familiar with the first chapters, he is ready to explore other technical skills or options of personal interest.

The book is based on a previous Spanish version that has been improved – thanks to the feedback gathered – and adapted to Minitab 16. It is a compilation of our experience of many years' teaching industrial statistics to undergraduate and graduate engineering and statistics students at Universitat Politècnica de Catalunya-Barcelona Tech; as well as experience gathered by training more that 600 Six Sigma Black Belts from different corporations. Our work as consultants in industrial statistics and quality improvement projects has formed the basis of deciding the content and has provided the basic material for many of the examples and cases presented. The data corresponding to examples and case studies presented are available on the publisher's website www.wiley.com/go/industrial_statistics_with_minitab.

We wish to express our thanks to our fellows at UPC: Josep Ginebra, Jan Graffelman, Alexandre Riba, Lourdes Rodero, Ignasi Solé and Moises Valls, without a doubt our best source of information; to Sandrine Santiago, ex-student and manager of CALETEC consultants, for her brilliant contributions; to our friend Guillermo de León, Professor at Universidad Veracruzana (Mexico), for many useful suggestions that have improved the clarity of many issues and to Lesly Acosta, PhD student and Assistant Professor at UPC, who has done a tremendous job translating from Spanish to English and providing valuable ideas. And finally, a special gratitude to Shubham Dixit from Aptara India for an excellent job correcting the manuscript.

Finally, we would be grateful if you let us know your comments and suggestions.

Pere Grima Cintas, Lluís Marco-Almagro
and Xavier Tort-Martorell Llabrés
Barcelona

Part One

INTRODUCTION AND GRAPHICAL TECHNIQUES

As much as 95% of all problems within a company can be solved using the seven basic tools.

K. Ishikawa

A good way to start learning how to use statistical software is to give an overview: see what it looks like, consider the different possibilities offered, enter some data, develop an idea on how to handle it and start doing some simple analysis. This is what we do with Minitab in Chapter 1.

The remaining chapters of Part 1 are devoted to graphical techniques. One might think that this is an easy subject, but when we are faced with some data – usually in a format and a structure that are not best suited for our purposes – it is not always easy to decide upon the best graph (or graphics) to convey the desired information. To make a good, informative graphic requires patience: producing different types and different versions of the most promising ones, subject knowledge, and mastering the software to make them. The examples presented here are aimed at presenting and practising Minitab using its many graphical capabilities, while developing practice in identifying the graphic type most suited to different situations.

The graphs presented are organized into three groups: to analyze variability, to split the data into categories to compare frequencies and to show the relationship between variables.

To analyze the variability of a data set we can use:

- Histograms. This is the graph of choice for visualizing dispersion. It is very useful to compare process output with product tolerances. And also very useful to stratify data – by machine, shift, operator, etc. – and compare it.

- Dotplots. This is similar to histograms and is especially recommended when you have few data. They are also very useful for comparing different situations by placing one diagram above another.

- Boxplots. These are very powerful and useful graphs. They have a box-shape showing the measures of position (quartiles), in addition they highlight the outliers. These plots are very intuitive and specially suited to compare a large number of groups.

To split the data into categories and show frequencies, there are:

- Bar Charts. These are similar to histograms but with a qualitative variable (gender, process, zone, etc.) or discrete (number of defects, age, etc.) on the horizontal axis instead of a continuous variable (weight, moisture, density, etc.). The bars can also be stratified.

- Pie Charts. These are widely used in commercial environments to show the size of the different parts of a whole. In general, their use is not recommended in technical environments.

- Pareto Diagram. This diagram is one of the most useful and widely used charts in quality management. They are a special case of bar charts and serve to show the parts of a problem, hoping to find the so-called Pareto principle: the vital few and the trivial many. As available resources are not usually sufficient to address all the causes at the same time, efforts should start and concentrate on the most important ones.

Cause-and-effect diagrams are a special case. Although they are not a graphical representation of data, we have included them alongside Pareto diagrams because they are two very typical quality improvement tools that often are used together. Naturally, Minitab does not help to identify possible causes (which is the important task) but facilitates the construction of the diagram.

To study the possible relationship between variables, graphs can be:

- Scatterplot. Minitab has many options and aids to produce, analyze and interpret this type of chart. There are especially useful the options to add a small dispersion (jitter) to avoid overlapping points and to use a tool (brush) to mark points and identify their origin.

- Time Series Plot. It can be seen as a particular case of scatterplot when the horizontal axis represents the order in which data is taken. It is very useful for analyzing the evolution of a variable over time.

- Graphics in three dimensions. Part 1 ends with a chapter on this type of graphics, for which Minitab presents many possibilities. Although they can be spectacular, frequently simpler graphics are better alternatives. They are very useful when working with response surface methods (Chapter 28).

As Ishikawa points out, a lot of the problems most frequently found in practice can be diagnosed and thus solved using the graphics that are covered in this first part. Naturally, for this to be true, quality data is needed. But this is another issue!

1

A First Look

1.1 Initial Screen

Main menu bar

Clicking on any option opens the corresponding submenus.

Toolbar containing action buttons

Placing the cursor over any of these icons a brief description is shown on its functionality.

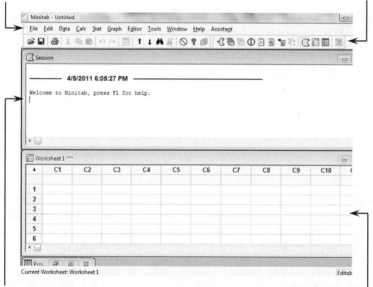

Session window

The results of any carried out analysis are displayed in this window. It may also be used to enter commands manually as an alternative to the use of the options in the main menu bar.

Worksheet

Very similar to a spreadsheet, with rows and columns. Columns are referred to by the labels C1, C2, ... just as shown, but may also be given a specific name, which can be written below C1, C2,...

Industrial Statistics with Minitab, First Edition. Pere Grima Cintas, Lluís Marco-Almagro and Xavier Tort-Martorell Llabrés.
© 2012 John Wiley & Sons, Ltd. Published 2012 by John Wiley & Sons, Ltd.

Usually, data are arranged in columns: each column represents a variable, and within a column, each row corresponds to an observation. In addition, values may be assigned to constants (K1, K2, ...) or matrices (M1, M2, ...). However, neither constants nor matrices appear in the worksheet.

The available submenus depend on which Minitab window is active (Session window, Worksheet, Graph, ...). Any window becomes active by just clicking on it.

1.2   Entering Data

Data can be introduced directly into the worksheet using the keyboard or retrieved from a previously stored file.

Example 1.1: Enter the data shown below. They represent the sales, in thousands of euros, in four different geographical zones during the first 5 months of the year.

	C1-T	C2	C3	C4	C5	C6	C7	C8
	Zone	January	February	March	April	May		
1	North	24	32	45	56	76		
2	South	62	34	23	6	23		
3	East	3	35	45	67	67		
4	West	34	78	23	44	58		
5								
6								

Worksheet 1 ***

The little arrow in the upper left corner of the worksheet indicates in which direction the cursor is moved when you enter an observation and hit the [Enter] key. By default, it points downwards; clicking on it, the arrow will point to the right meaning that data would be entered row-wise.

As mentioned, by default columns are named C1, C2, ..., but the row just below may be used to introduce specific names to those columns; this is highly recommended. In the previous worksheet, the heading given to column 1 is C1-T meaning that this column contains text data.

It is possible to create a new column as the sum of two others, for example, C3 = C1+C2, in such a way that changing C1 or C2 will also change C3 (see 'Data operations' below). Note that it is only possible to link columns but not individual cells.

1.3   Saving Data: Worksheets and Projects

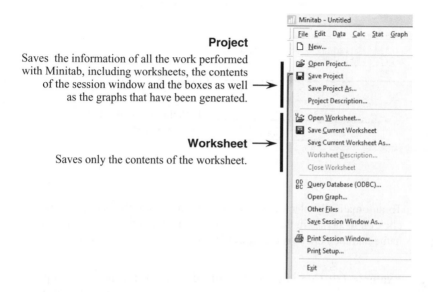

Project

Saves the information of all the work performed with Minitab, including worksheets, the contents of the session window and the boxes as well as the graphs that have been generated. ⟶

Worksheet ⟶

Saves only the contents of the worksheet.

The used built-in Minitab extensions are .MTW for a worksheet and .MPJ for a project.

A file can only be retrieved in the same way it was previously saved. If it was stored as a **worksheet**, it can only be retrieved as such. The same is valid for **projects**.

Minitab interacts very well with Excel; an Excel spreadsheet can be imported using the option **Open Worksheet** (or copy and paste from Excel to Minitab).

1.4   Data Operations: An Introduction

When starting to work with Minitab, the easiest way to carry out data calculations is by using the option **Calc** > **Calculator**. The following dialog box shows how to create a new variable, named Total, as the sum of the entries in columns C2 to C6. Notice that this new variable is placed in the first empty column found in the worksheet.

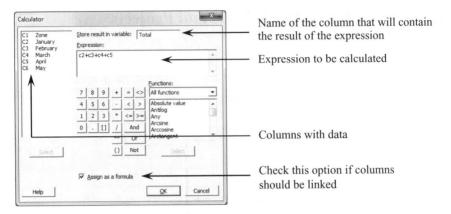

Name of the column that will contain the result of the expression

Expression to be calculated

Columns with data

Check this option if columns should be linked

To automatically place a column in the box where the cursor is, double-click on its name (in the left variable list) or alternatively click once on its name and then click on the **Select** button.

Example 1.2: Consider the data already used in Example 1.1. Suppose that there is a commercial team per each geographical zone. It is decided that 10% of each zone sales amount, collected during the first four months of the year, should be split among the team-members of each zone. Carry out the appropriate operation and store the quantity corresponding to each commercial team in column C8.

To achieve this, use the menu option **Calc > Calculator,** then introduce C8 in box **Store result in variable** and finally write $0.1*(C2+C3+C4+C5)$ in the **Expression** box.

Another way to carry out operations is through the menus **Calc > Column Statistics** or **Calc > Row Statistics**.

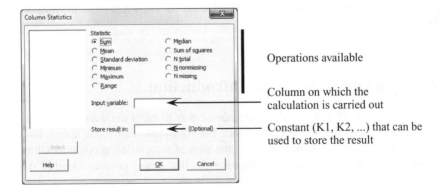

Operations available

Column on which the calculation is carried out

Constant (K1, K2, ...) that can be used to store the result

Example 1.3: Once again, consider the data of Example 1.1. Herein, the objective is to determine, using both options: **Calculator** and **Row Statistics**, the average monthly consumption per geographical zone.

To achieve this do: **Calc > Calculator**, then **Store result in variable**: C7 (for example) and finally **Expression**: (C2+C3+C4+C5+C6)/5.

Alternatively, use: **Calc>Row Statistics** in the way shown below.

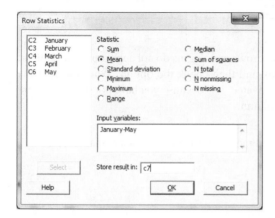

If C2 C6 is entered, we refer to these two columns entries.
If C2-C6 is entered, we refer to all entries in columns C2 to C6: C2, C3, C4, C5, and C6.

To select a set of consecutive variables in a single action: first click on **Input variables** (the variables list box is displayed on the left), then select the relevant columns by first clicking on the first variable and then drag the cursor until the last one; finally click on **Select.**

1.5   Deleting and Inserting Columns and Rows

Delete a column	Click on the name of the column (by default the names are: C1, C2, ...) and press [Del] key.
Delete a row	Click on the row number and press [Del] key.
Delete a cell	Place the cursor in the cell and press [Del] key.
Insert a row	Click on the row number, above which you want to insert the new row, and click on the icon **Insert Row**.

Insert a column   Click on the name of the column (C1, C2, ...), to the left of which you want to insert the new one, and then click on the icon **Insert Column**.

 If the icons needed to insert rows or columns do not appear on your toolbar, go to **Options** > **Toolbar** and then check the option: **Worksheet.** If the icons appear but are not activated, it is because you have not selected where to insert the row or the column.

 In Appendix 2 you will find more information on data management.

1.6   First Statistical Analyses

 The file named DETERGENT.MTW contains in column C1 the weight (in grams) of 500 packages of washing powder whose nominal weight is 4 kg. Column C2 indicates which of the two available filling lines was used.

To study the distribution of the variable weight stratified by the categorical variable filling line, do the following:

Stat > **Basic Statistics** > **Display Descriptive Statistics**

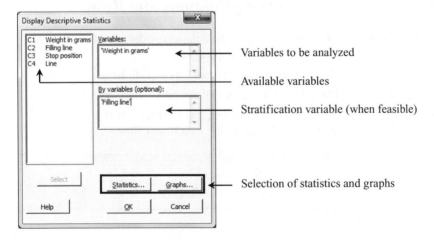

Recall that if you double-click on a variable's name (in the variable list box on the left), this variable will be automatically moved into the box where the cursor is. You may also click just once on the variable's name and then click on **Select**.

Without changing the default options, the following output will appear in the Session window:

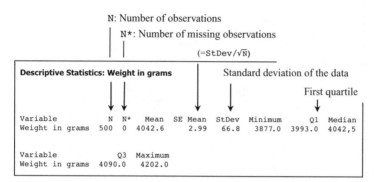

N: Number of observations

N*: Number of missing observations

(=StDev/√N̄)

Descriptive Statistics: Weight in grams

Standard deviation of the data

First quartile

```
Variable          N   N*     Mean   SE Mean   StDev   Minimum      Q1   Median
Weight in grams  500   0   4042.6      2.99    66.8    3877.0  3993.0   4042,5

Variable             Q3   Maximum
Weight in grams  4090.0    4202.0
```

Clicking on **Statistics** opens the following dialog box displaying a range of possible statistics to be computed. The statistics chosen by default are the ones with a checkmark.

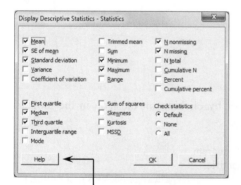

Clicking on **Help** provides information on the meaning of each of these statistics. For example: **Trimmed mean** is the arithmetic mean of the data after eliminating the 5% largest and the 5% lowest data points. It is a measure of central tendency less sensitive to outliers than the arithmetic mean.

The option **Graphs** allows you to choose among the following graphs list (by default, none will be drawn):

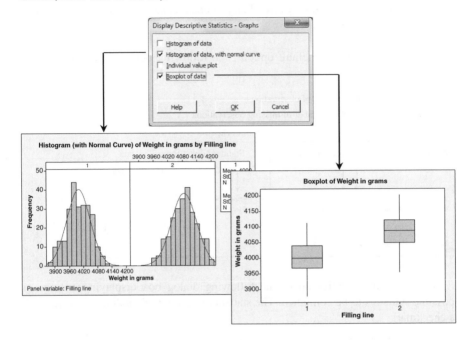

1.7   Getting Help

1.7.1   The Main Help Menu

It can be accessed via the main menu or by clicking on the icon in the toolbar. Alternatively, you may also hit the **F1** key.

Keyword search. For example start writing: histogram.
After writing the first three letters Minitab identifies
the word that we are looking for.   Hypertext guide. VERY COMPLETE

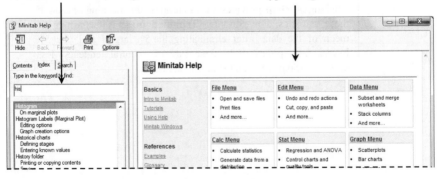

Double-clicking on the chosen option, information is shown on how to draw a histogram using the main menu. It is also always interesting to check the links on the top.

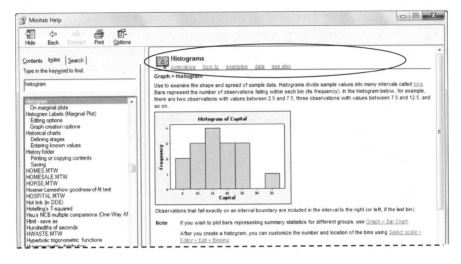

1.7.2   Contextual Help

If you are carrying out a statistical analysis and need help on a specific topic, there is always a **Help** option in each dialog box. Click on it to open a new window that contains the requested information. For example, for **Calc > Calculator:**

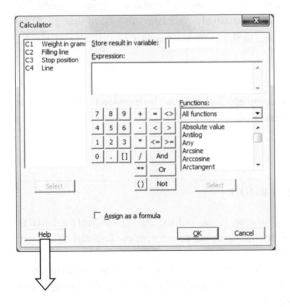

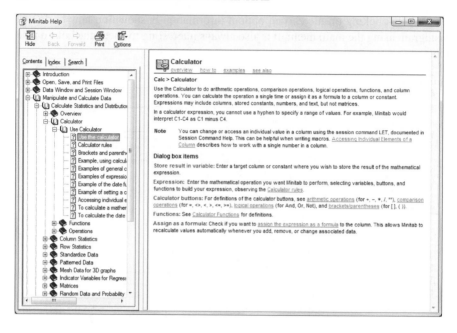

1.7.3   StatGuide™

Right-clicking over the results obtained via a statistical analysis (either in a Session window or in a Graph window) and then clicking on **StatGuide** opens a guide that helps you to interpret such results. Alternatively, you may also open that guide by either clicking on the icon located in the toolbar or by hitting [Shift] + [F1].

Minitab's help is very complete and very well organized. This makes it easy to find whatever you need.

1.8   Personal Configuration

Almost everything, from the appearance (font, colors, etc.) to aspects related to configuration, can be customized according to your preferences. The way to achieve this is through: **Tools** > **Options**

For example, if you do not want Minitab to prompt you whether you want to save a graph every time you close a Graph window, choose **Graphs** > **Graph Management** and activate the corresponding option.

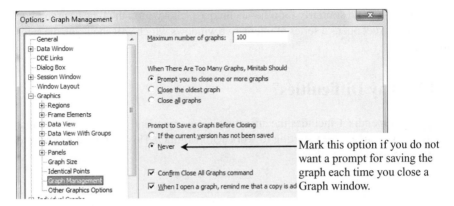

Mark this option if you do not want a prompt for saving the graph each time you close a Graph window.

 To restore the configuration's default settings, you need to execute the file rmd.exe (Restore Minitab Defaults) found in the directory where Minitab was installed.

Once you are familiar with the Minitab work environment, you may want to consult Appendix 3 for more detailed information on how to customize Minitab.

1.9   Assistant

Minitab Release 16 also incorporates an **Assistant** (just to the right of **Help** in the main menu) which may guide you in the application of some specific techniques and which may be useful to you as an initial reference.

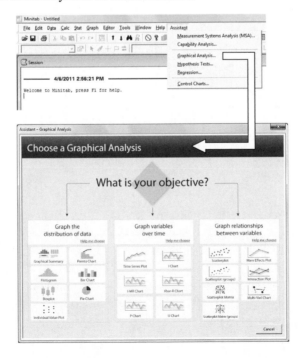

In this case, and similarly for other techniques, clicking on the type of graph you want to construct opens a series of interactive windows which will show you the steps to follow.

1.10   Any Difficulties?

 Appendix 1 includes the answers to some frequently asked questions that may arise when starting to work with Minitab. This is a good moment to have a look at them.

2

Graphics for Univariate Data

2.1 File 'PULSE'

 An experiment is conducted in a class with 92 students, where initially each student registers his or her own height, weight, gender, smoking preference, usual physical activity level and resting pulse rate. Then, each student flips a coin and those who get a face coming up must run in place for one minute. Finally, after that time, all measure and register their new pulse rates.

The file named PULSE.MTW, included in the Minitab sample data folder, will be used throughout this chapter to illustrate the different type of graphs that can be constructed. This file is organized as follows:

Column	Variable	Label
C1	Pulse1	First pulse rate of the 92 students
C2	Pulse2	Second pulse rate
C3	Ran	1=ran in place; 2=did not run in place
C4	Smokes	1=smokes regularly; 2=does not smoke regularly
C5	Sex	1=male; 2=female
C6	Height	Height (in inches)
C7	Weight	Weight (in pounds)
C8	Activity	Usual level of physical activity: 1=slight; 2=moderate; 3=a lot

To retrieve the worksheet file do: **File > Open worksheet** and click over the button **Look in Minitab Sample Data folder**.

Industrial Statistics with Minitab, First Edition. Pere Grima Cintas, Lluís Marco-Almagro and Xavier Tort-Martorell Llabrés.
© 2012 John Wiley & Sons, Ltd. Published 2012 by John Wiley & Sons, Ltd.

Information about the content of the different files included in the Minitab sample data folder can be obtained from the **Help** menu. Specifically, for information of the file 'Pulse' you can do: **Help** > **help** and write **Pulse** in the index.

2.2   Histograms

Graph > **Histogram**

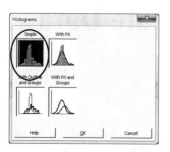

Visual gallery from which to choose a type of histogram

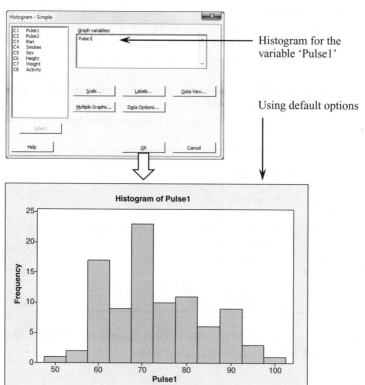

Histogram for the variable 'Pulse1'

Using default options

2.3   Changing the Appearance of Histograms

2.3.1   Changes in the Scale of the Horizontal Axis

Double-clicking over any scale value on the horizontal axis, a dialog box pops up. The default values are then changed as follows:

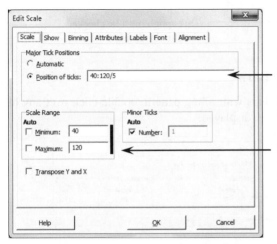

New values used for the scale go here. First, disable the option that appears by default (**Automatic**) and then indicate that these new values are 40 to 120 with increments of 5

Minimum and maximum scale values (the default values have been substituted)

2.3.2   Changes in the Vertical Axis

In this axis the only change with respect to default settings that we make consists in introducing four marks between Y-scale values (these marks are called minor ticks and their use aims to identify values over the scale without indicating the corresponding numbers). To do this:

Double-click over any vertical scale value to display the same dialog box mentioned above:

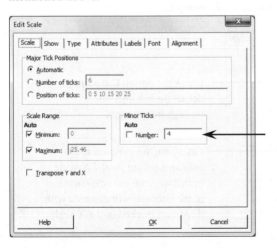

Only change this value ("Number of Minor Ticks in the scale")

Also go to the tab **Show** to indicate that the minor ticks must be shown.

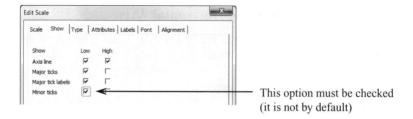

This option must be checked
(it is not by default)

2.3.3   Appearance of the Bars

If, for example, we do not want bars with filling pattern, double-click over any of them. The following window is then displayed:

Check this option

Choose the option: without filling pattern, which is the one positioned just above the default one.

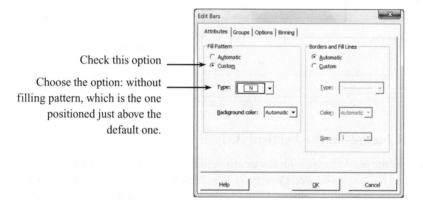

2.3.4   Defining Intervals on which Bars are Located

Double click over any bar and click on the **Binning** tab.

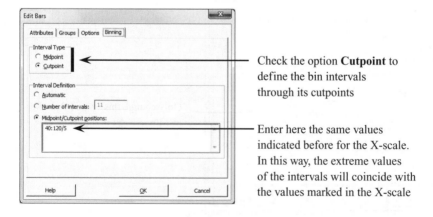

Check the option **Cutpoint** to define the bin intervals through its cutpoints

Enter here the same values indicated before for the X-scale. In this way, the extreme values of the intervals will coincide with the values marked in the X-scale

2.3.5   Appearance of the Graph Window

We will remove the gray border that appears around the histogram. Double-click over this region (outside the border of the histogram):

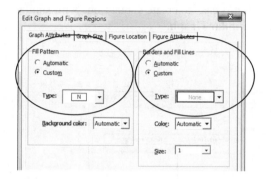

Change default options to eliminate the background color (**Fill Pattern**) and the exterior line (**Borders and Fill Lines**)

2.3.6   Changes in the Proportions of a Graph

Sometimes it is convenient to change the default proportions of a graph. To do so, open the previous dialog box and click on the **Graph Size** tab.

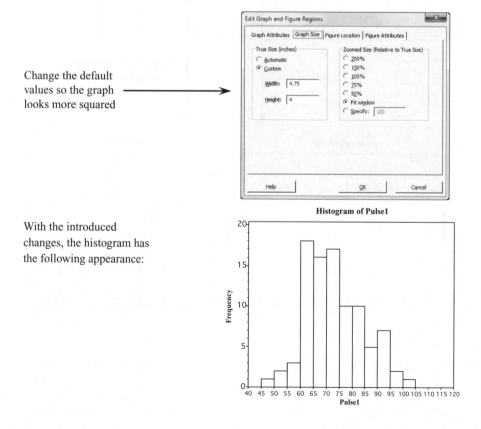

Change the default values so the graph looks more squared

With the introduced changes, the histogram has the following appearance:

The edition and modification of graphs is easy and intuitive. It basically consists on double-clicking over the element to be changed.

Once a graph has been customized, it is easy to create another one with similar characteristics using: **Editor > Make Similar Graph**.

Editor > Make Similar Graph

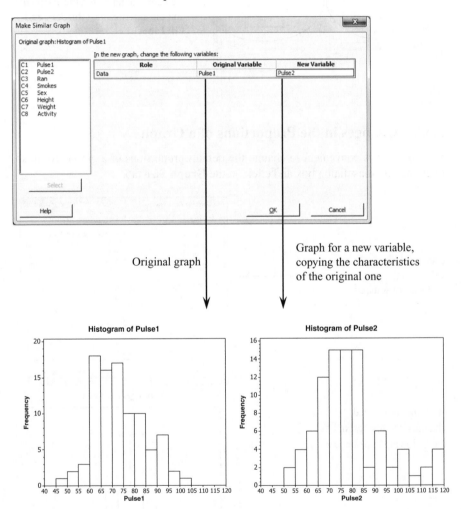

Original graph

Graph for a new variable, copying the characteristics of the original one

2.4   Histograms for Various Data Sets

Using the option **Calculator** located within the **Calc** menu, we can create the column named 'Increment' that defines the difference between the final pulse rate and the initial pulse rate as follow.

Calc > Calculator

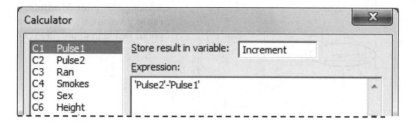

A comparison of pulse increments, depending on whether the student has run or not, can be done through their histograms, as follows.

Graph > Histogram: with Outline and Groups
Putting 'Increment' as the variable to represent and 'ran' as a categorical variable to form the groups, we obtain (the display shows a different color for each histogram):

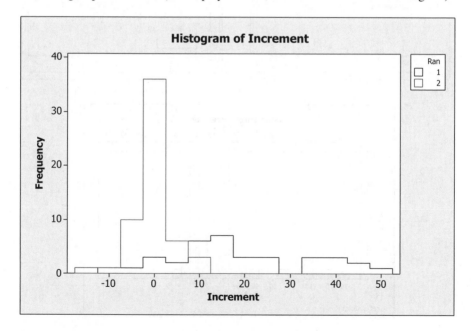

Another option with a more clear result is **Graph > Histogram: Simple**.

Graph > Histogram: Simple

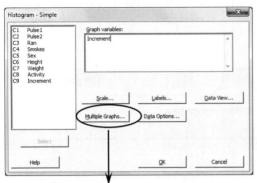

Choosing the option **Multiple graphs** gives access to additional options that help to provide very clear outputs

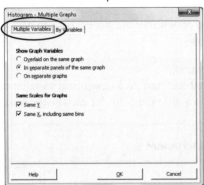

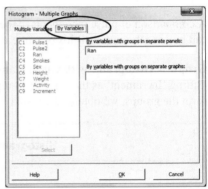

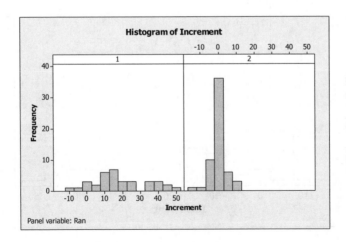

To have a list of all available graphs and worksheets click on **Window** in the main menu. All worksheets and graphs created and not deleted appear at the end.

2.5   Dotplots

Graph > Dotplot
Use the available graphical options to: first, compare the pulse rate increments according to the running condition and then do the same but also identifying the points by gender.

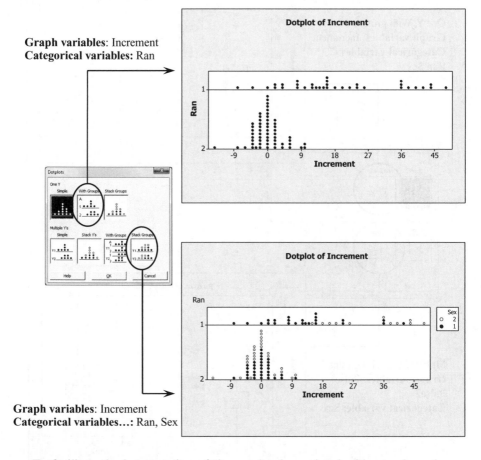

Graph variables: Increment
Categorical variables: Ran

Graph variables: Increment
Categorical variables…: Ran, Sex

To facilitate the interpretation of the graph when printed without colors, the default type of symbol used for females can be changed (sex=2). The process to carry out such change is: 1) Click over any point; 2) Click over any point of the type that is to be changed; 3) Double-click over the same point 4) Choose the desired type of point in the dialog box that appears.

The appearance of a dotplot diagram can be changed by editing it in just the same way as with histograms. Indeed, the graphical options are the same ones used previously.

2.6   Boxplots

Graph > Boxplot

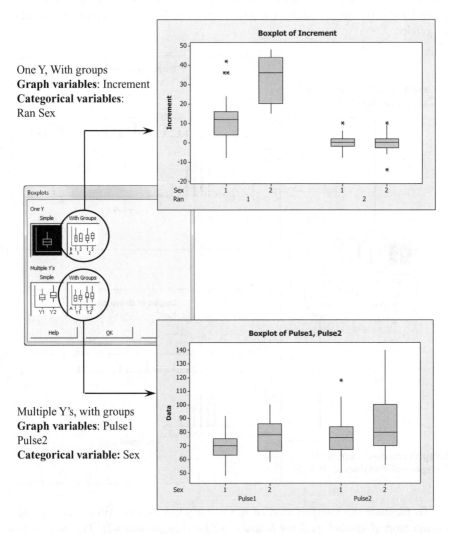

One Y, With groups
Graph variables: Increment
Categorical variables:
Ran Sex

Multiple Y's, with groups
Graph variables: Pulse1
Pulse2
Categorical variable: Sex

Using default options, the usual appearance of a boxplot is: a box delimiting the interquartile range (IQR) with interior line indicating the median, whiskers as far as the last observation within the region delimited by the quartiles $+1.5$ IQR and values located further away from this region identified by asterisks.

2.7   Bar Diagrams

Graph > Bar Chart

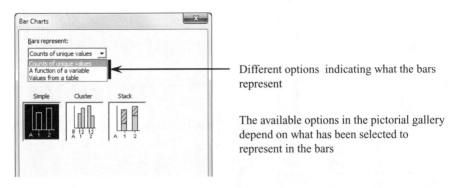

Different options indicating what the bars represent

The available options in the pictorial gallery depend on what has been selected to represent in the bars

For example: We could represent the number of men and women according to their physical activity level.

Graph > Bar Chart: Counts of Unique Values, Stack
Categorical variables: **Activity Sex**

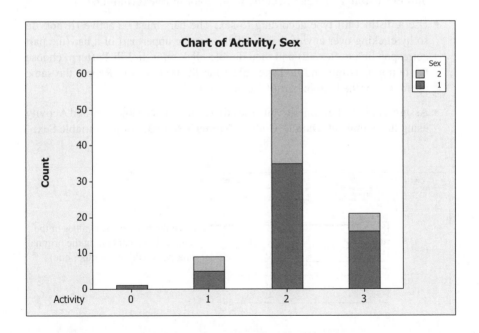

Transform the previous graph that uses the default options, for the next one:

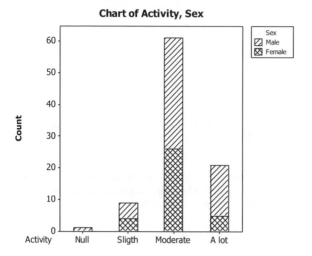

The carried out procedure to change the appearance of the bar chart is:

- Remove the color from the bars: Double-click over any bar to open up a dialog box (headed by **Edit bars**, tab: **Atributes**) and in **Fill Pattern** choose **the option Custom** and then select the white color in **Background color**.

- Use a distinct fill type according to sex. The bars must be active (if not, do so by clicking over any of them). Click over the upper part of a bar (the part corresponding to Sex=2) and then double-click on it. In **Fill Pattern,** choose the option **Custom** and in **Type** select the **fill type** to use. Repeat the same operation for the bottom part of the bar.

- Assign names to the numeric values used to codify the variables Sex and Activity using the command: **Data > Code > Numeric to text**. For the variable Sex:

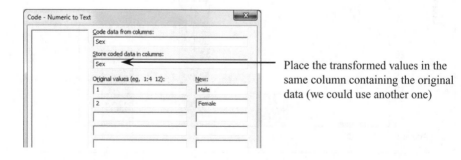

Place the transformed values in the same column containing the original data (we could use another one)

Likewise, the values of the variable activity can be transformed substituting them for their corresponding labels.

There is no need to construct another graph after changing the values that identify the variables Sex and Activity. It is enough to right-click on any point in the graph and check the option: **Update the graph automatically.**

- Once the codes of the variable activity are transformed by their labels (Null, Slight, Moderate, A lot), sort all columns in increasing order by the variable activity. Then, in the worksheet, select the column **Activity** (clicking over **C8-T**) and go to: **Editor** > **Column** > **Value Order**: select the option **User-specified order**, define the order and click on the button **Add order.**

- Remove the grey border: Double-click on any point in the border. In **Attributes** remove the background color (**Fill Pattern, Custom**) and the exterior line (**Borders and Fill Lines, Custom**) in a similar way as done for the histogram.

- Change the dimensions: Double-click over the exterior part of the graph, click on the tab **Graph Size. In True size, Customized** enter **Width**: 4.75, **Height**: 4 (if units are inches, write other reasonable values if units are millimeters).

2.8   Pie Charts

Graph > **Pie Chart**

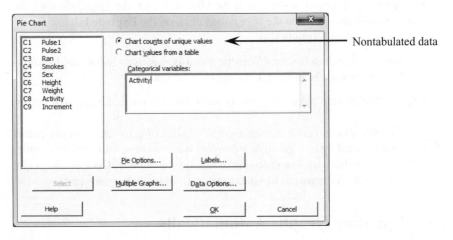

To obtain this appearance we have used fill types instead of colors as done before with the bar charts. Also the options described below have been selected.

Pie Chart of Activity

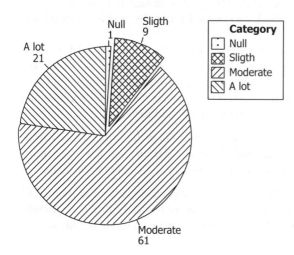

To obtain this appearance we have used fill types instead of colors as done before with the bar charts. Also the options described below have been selected.

- Explode or pull away a slice: Edit the slice (click on the graph, click on the slice and double-click on the slice), then click on the **Explode tab** and finally check the option **Explode slice**.

- Name and frequency of slices: With the graph as an active window, do: **Editor** > **Add** > **Slice Labels** and check the appropriate options.

- Graph background color and graph borders: Edit the graph, **Graph Attributes.**

 Notice that to edit a part of a graph (a slice of a pie chart, or the points corresponding to a group in a dotplot) it is not enough to click or double-click on that part. The procedure is always the same: 1) Click on the graph; 2) Click on the part to be edited and 3) Double-click on this part.

2.9   Updating Graphs Automatically

Right-clicking on a graph the following menu appears:

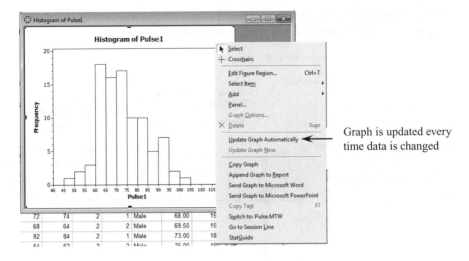

Graph is updated every time data is changed

If the automatic update is activated, the graph changes as the data used to construct it changes (either by adding, modifying or removing some data).

When any value in the worksheet is changed the graph can be updated at that moment (**Update graph now**) but it will not be updated automatically when further changes are done.

The icon that appears in the left superior corner of the graph indicates whether it has been updated or not, according to the next code:

Icon/Color	Meaning
Green	The graph corresponds to the values in the worksheet. It appears the first time that the graph is constructed, or when it has been updated (automatically or manually) after modifying the data.
Yellow	The graph does not correspond to the data, since it has been modified and the graph has not been updated.
Red	The graph cannot be updated because the changes introduced in the data make it impossible. For example: some of the values used for the construction of a bivariate diagram have been removed and the two columns are not of equal length.
White	The data has changed, but it is not possible to update that type of graph. For example: Pareto diagrams.

2.10   Adding Text or Figures to a Graph

Choose as an active window the graph to be modified. Using:

Editor > Anotation > Graph Annotation Tools
a bar appears allowing the addition of text and figures to a graph.

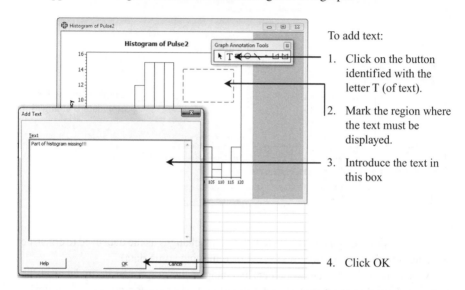

To add text:

1. Click on the button identified with the letter T (of text).

2. Mark the region where the text must be displayed.

3. Introduce the text in this box

4. Click OK

The ellipse and the line (that once edited becomes an arrow) are drawn by clicking over the corresponding icons shown in the menu. All introduced elements can be edited (by double-clicking) and modified (increasing the font size of the text, changing the type of line from a continuous to a discontinuous line . . .) as needed.

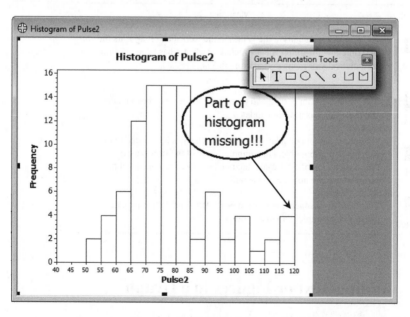

3

Pareto Charts and Cause–Effect Diagrams

3.1   File 'DETERGENT'

Consider again the file DETERGENT.MTW of Chapter 1, containing data of a company that produces washing powder for washing machines. Through an inspection, carried out during the last quarter of the year 2009, the company detected very frequent stops in the two available production lines. The production process disposes of an automatic register system which records, every time there is a stop, the position of the stop (riveter, belt, feeding, filling, sealing, palletization) and the line where it occurs (two available lines). The Columns C3 and C4 of file DETERGENT.MTW contain the position of the stops and the production line where the stops occur, respectively.

Industrial Statistics with Minitab, First Edition. Pere Grima Cintas, Lluís Marco-Almagro and Xavier Tort-Martorell Llabrés.
© 2012 John Wiley & Sons, Ltd. Published 2012 by John Wiley & Sons, Ltd.

3.2   Pareto Charts

Stat > Quality Tools > Pareto Chart...

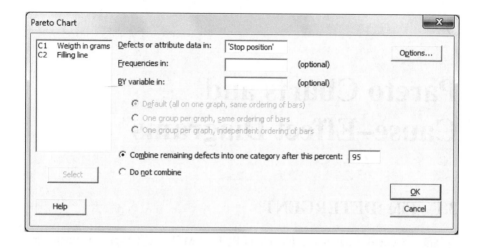

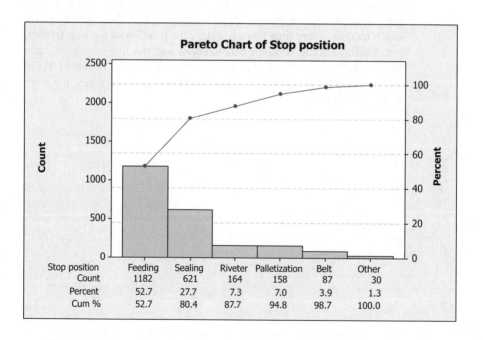

Using all default options, introduce the variable 'Line' in '**BY variable in**' to construct a Pareto chart of the variable 'Stop position' stratified by the production 'line' variable, as shown below:

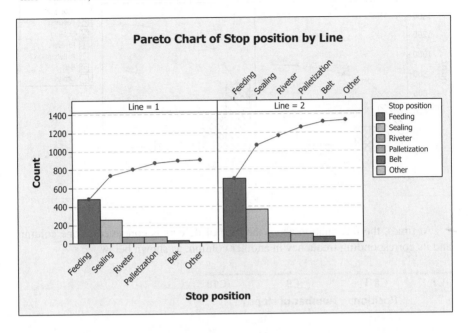

3.2.1   Changing Appearance of Pareto Charts

To change the bar colors (something highly recommended if you want to produce black and white graphical outputs), follow the next steps:

- Delete the colors (that is, choose the white color for all bars): Double-click on any bar (all bars are selected at once), then in **Fill Pattern** choose the **Custom** option and finally select the desired color (in this case white) **in Background color**.

- Add the line patterns: once all bars are selected (done as indicated in the previous step), click on any of them (so only this one is selected), and then double-click on it (a dialog box for edition appears). In **Fill Pattern**, first choose the **Custom** option and then select the desired **Type** pattern. Repeat the same procedure with all other bars.

- Remove the grey border: Double-click anywhere in the grey graph region to open a dialog box. in **Fill Pattern** and **Borders and Fill Lines**, first choose the **Custom** option and then select the appropriate options to produce a graph with the following appearance:

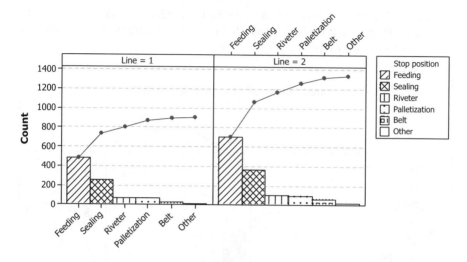

↝ At times, the information about the type of defects is contained in one column and its corresponding frequency in another column, as shown below:

C7	C8-T	C9	C10
	Position	Number of stops	
	Feeding	1182	
	Sealing	621	
	Riveter	164	
	Palletization	158	
	Belt	87	
	Filling	30	

In such cases, the dialog box would have a different look:

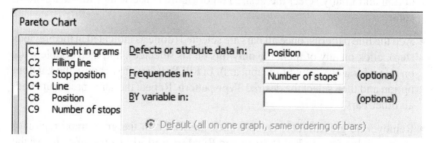

The resulting plot, as expected, would coincide with the nonstratified case (the first constructed Pareto chart corresponding to the variable 'Stop position').

 Stratification is only possible if the data are not-grouped, but you can identify to which group each data point belongs. On the contrary, stratification of grouped data presented in tabular form is not possible.

Notice that, by default, Minitab assigns names to each bar until they sum up to 95% of the total; the remaining bars are grouped under the category '**Other**'; this value of 95% can be changed.

3.3   Cause-and-Effect Diagrams

Stat > Quality tools > Cause-and-Effect
First, the potential causes must be entered in the worksheet. For example:

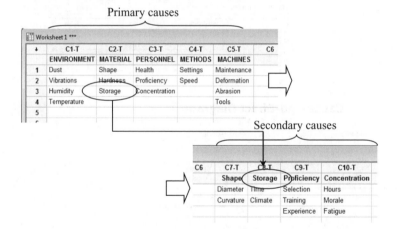

Once all potential causes are introduced in the worksheet, go to: **Stat** > **Quality Tools** > **Cause and Effect**...

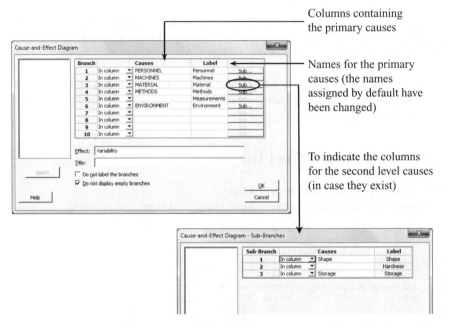

Columns containing the primary causes

Names for the primary causes (the names assigned by default have been changed)

To indicate the columns for the second level causes (in case they exist)

Cause-and-Effect Diagram

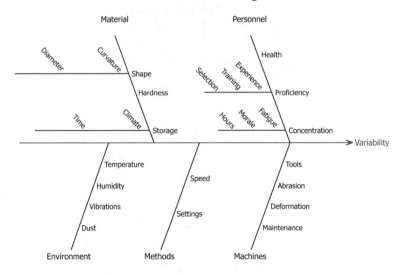

 Concerning the cause-and-effect diagrams, Minitab is only useful to present these causes in a clear way, but obviously it does not contribute anything to the list of potential causes nor to the analysis of which are related to the effect under study.

4

Scatterplots

4.1 File 'Pulse'

 Consider the data in file PULSE.MTW already described in Chapter 2.

Graph > Scatterplot

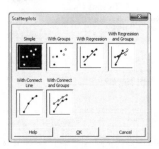 This initial dialog box displays a pictorial gallery that allows you to choose among different type of scatterplots; the most commonly used are: **Simple** and **With Groups**.

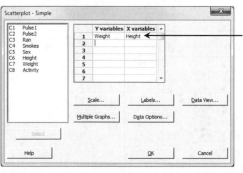 In **X variables,** place the columns containing the values of X (horizontal axis) and in **Y variables** the ones containing the values of Y (vertical axis)

Industrial Statistics with Minitab, First Edition. Pere Grima Cintas, Lluís Marco-Almagro and Xavier Tort-Martorell Llabrés.
© 2012 John Wiley & Sons, Ltd. Published 2012 by John Wiley & Sons, Ltd.

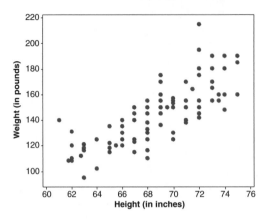

To get this plot: first, change the name of the columns so that the units are displayed on the axes. Then, eliminate the border and the background color. Finally, modify the graph dimensions to make it more squared.

4.2   Stratification

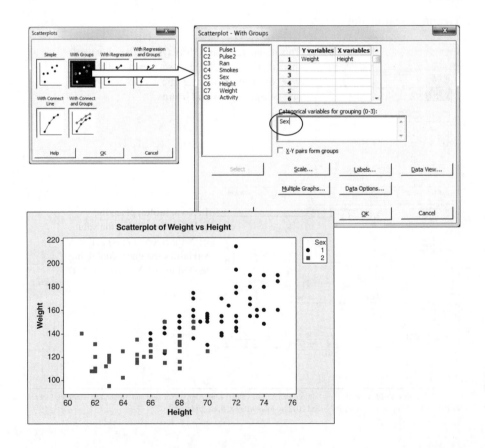

By default, Minitab distinguishes the groups (in this case, Sex = 1 or Sex=2) by different colors and symbols as shown above. However, when the graph is printed in black and white, the observed difference is not very clear.

If you want to change the type of symbol (and also the color and size) used for each group, follow the next steps:

First, click on any of the data points; this will select all of them. Second, click again on the same data point; this will now select solely the points of a group. Finally, double-click on the same data point to open a dialog box that will allow you to change the color, symbol and size for all points in that group.

Graph > Scatterplot. . . : With Groups is not the only way to get a stratified scatterplot. For instance, if you have drawn a scatterplot from **Graph > Scatterplot . . .: Simple**, you can also add a stratification variable by first double-clicking on any of the data points and then choosing the **Groups** tab in the dialog box that appears, as shown below:

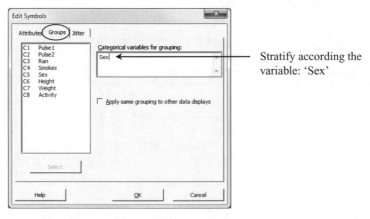

Stratify according the variable: 'Sex'

Notice that the result is the same as the one obtained when stratification is performed directly using the commands: **Graph > Scatterplot. . . : With Groups**.

4.3   Identifying Points on a Graph

Consider a new file, CARS.MTW, to demonstrate Minitab possibilities to identify points on graphs.

 This file contains 14 columns corresponding to different characteristics of 236 cars; the data is taken from the website of the Spanish Royal Automobile Club (*Real Automóvil Club de España*) in January 2011. This file is organized as follows:

Column	Variable
C1	Maker
C2	Model
C3	Price (in Euros)
C4	Number of cylinders
C5	Displacement (cc)
C6	Power (CV)
C7	Length (cm)
C8	Width (cm)
C9	Height (cm)
C10	Luggage capacity (litres)
C11	Weight (kg)
C12	Consumption (litrs/100 km)
C13	Maximum speed (km/h)
C14	Acceleration (seconds going from 0 to 100 km/h)

Plotting price(Y) with respect to power(X):

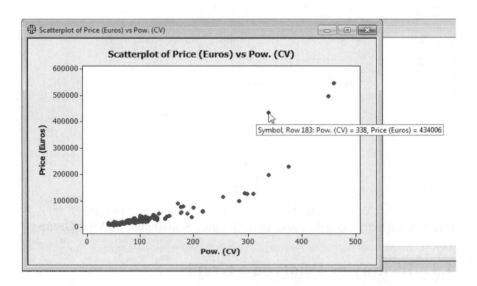

A simple way to identify cars with high prices in relation to their power consists on placing the cursor over the dots corresponding to observed high prices. Indeed, if you do so, after a while a small information box appears containing the row number corresponding to the data point together with its respective Y and X values.

Alternatively, you can identify points on a graph using: **Editor > Brush**. This will open up a small box containing the row-numbers corresponding to the marked data points. It is possible to mark a single data point (by clicking on it) or several at once (by enclosing them inside a rectangle drawn dragging the mouse).

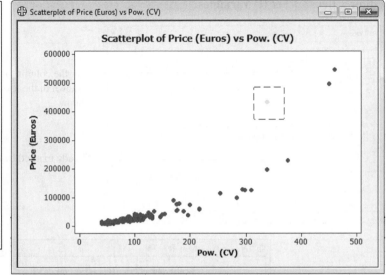

 A faster way to activate the **Brush** option is to do so from the main menu, using: **Tools > Toolbars** and then choosing the option **Graph Editing**. A simplest alternative would be to right-click anywhere on the main menu and then choose the option **Graph Editing**.

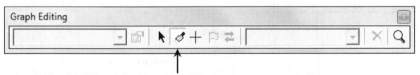

Click here to activate the Brush option on
the active graph window.

Once both the **Brush** option and the graph window are activated, clicking on **Editor** in the main menu shows the following Editor options:

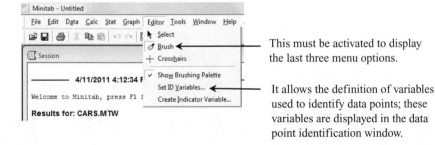

This must be activated to display
the last three menu options.

It allows the definition of variables
used to identify data points; these
variables are displayed in the data
point identification window.

Editor > Brush > Set ID Variables

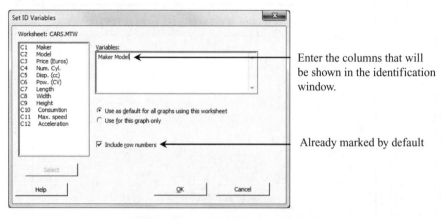

Enter the columns that will
be shown in the identification
window.

Already marked by default

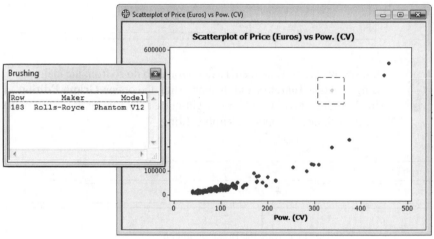

If several graph windows exist and in all of them the **Brush** option is active, the action of brushing points in one graph automatically does the same across graphs.

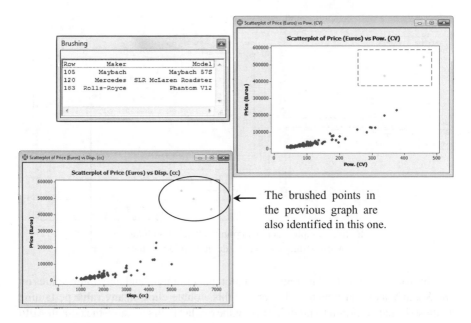

The brushed points in the previous graph are also identified in this one.

 To change the default color of the selected data points, go to: **Tools > Options > Graphics > Data View > Symbol:** choose the color in **Brushed Symbols.** This set up stays from session to session.

It is possible to label data points on a graph. To achieve this, depart from the initial dialog box, hit the **Labels** option, then click on the **Data Labels** tab.

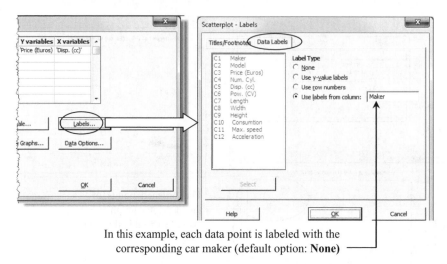

In this example, each data point is labeled with the corresponding car maker (default option: **None**)

To make a zoom of a data region of a plot, the minimum and maximum values on the X and Y axes must change. To achieve this, double-click on any value pertaining to the axis that you want to modify, then in the **Scale** tab first disable the default **Auto** option and then enter the new values in **Scale Range**.

X axis   Y axis

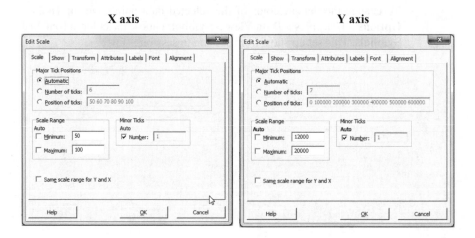

For the 'price' variable, set the minimum and maximum values of 12000 and 20000, respectively. Likewise, for the 'power' variable, set the minimum and maximum values of 50 and 100 CV respectively, to obtain:

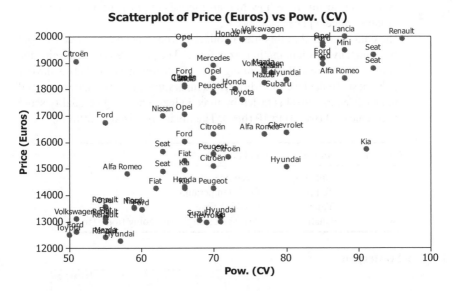

To select a label, click on it and after a while make a second click. In this way, you can change the label's size and location, though in the presence of many points overlapping of labels is inevitable.

4.4   Using the 'Crosshairs' Option

Use the **Crosshairs** option to get the X–Y coordinates of any data point on the graph. To activate it, go to: **Editor > Crosshairs** or click the icon fx in the **Graph Editing** toolbar. Notice that the cursor becomes a cross that can be moved over the graph and that the coordinates are shown in the upper left corner of the graph.

Coordinates of the data point marked by the cross

The cross moves over the graph data region as you move the mouse

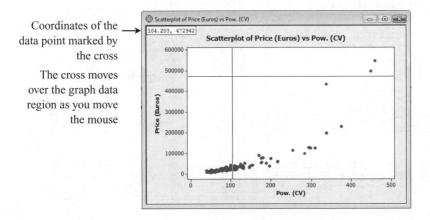

4.5   Scatterplots with Panels

Consider the data in file REHEAT.MTW, included in the Minitab sample data folder, to illustrate how to construct scatterplots stratified by a third variable called a panel variable.

The data contained in this file is collected from an experiment carried out in a company specializing in frozen foods. The purpose is to determine the oven temperature and the cooking time of a frozen meal so that its flavour is the best. To achieve this, different temperature and time values are used and each sample meal is tasted by three judges who assign a quality score on a scale of 0 (worst) to 10 (best). The file is organized as follows:

Column	Variable	Label
C1	Operator	Oven operator
C2	Temp	Oven temperature
C3	Time	Cooking time
C4	Quality	Average tasting score, going from 0 to 10

Graph > Scatterplot >

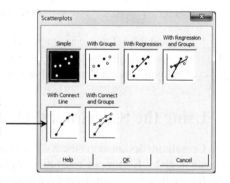

Choose **With Connect Line** to get a scatterplot with dots connected by lines. This allows a better visualization of the trends.

Choose **Quality** as **Y** variable and **Time** as **X** variable.

In the tab **By Variables**, introduce **Temp** as indicated

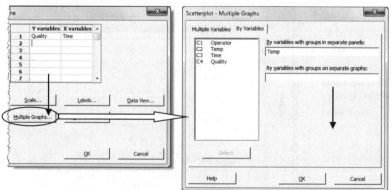

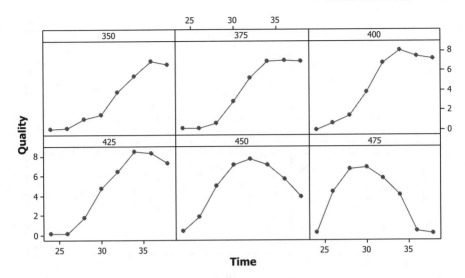

This plot shows a scatterplot of the Quality variable in relation to the time variable, for each temperature value used in the experiment. Observe that for low oven temperature values (the three scatterplots above) the best quality measures are obtained at very high values of time. On the contrary, with higher temperature values (bottom plots) best quality measures are already observed at much lower time values. Thus, with high temperature values, keeping a frozen meal too long in the oven is counterproductive because its quality degrades a lot (probably because it burns). Notice that the best quality scores are obtained at 425 °C and about 34 minutes.

You can change the appearance of a graph with panels. To do so, go to: **Editor > Panel**. . . and then click on the **Options** tab.

When having adjoining panels, you can choose to alternate the side on which axes labels appeara (the default) or choose to place all on the same side.

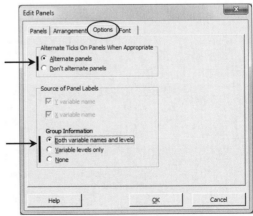

You can choose to display every panel with the name and value of the variable, only its value or none of them

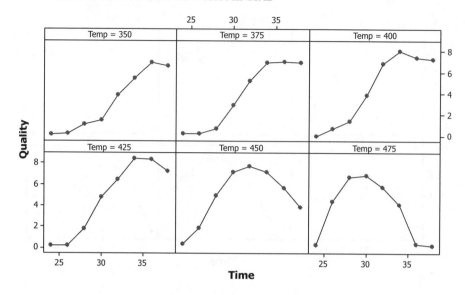

4.6   Scatterplots with Marginal Graphs

  Consider again the file CARS.MTW.

Graph > Marginal Plot

You can choose any of the three marginal plots options departing from the initial dialog box and then choosing the X and Y variables in the usual way.

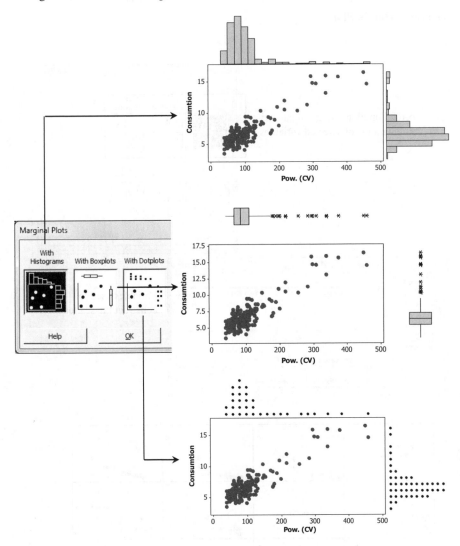

4.7   Creating an Array of Scatterplots

4.7.1   Square Matrix

Graph > Matrix Plot

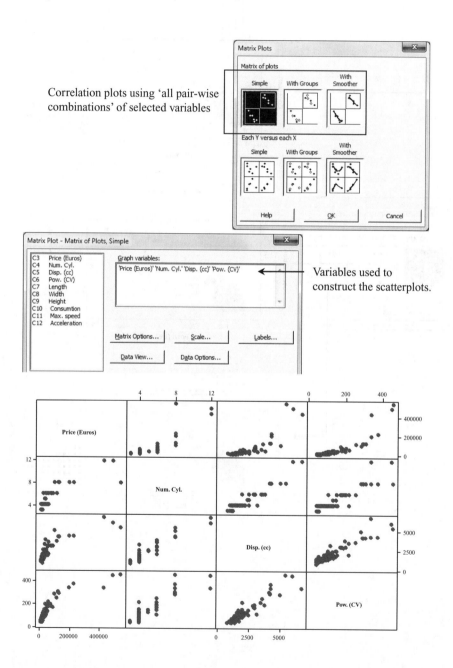

Correlation plots using 'all pair-wise combinations' of selected variables

Variables used to construct the scatterplots.

The use of too many variables produces plots difficult to interpret. Thus, any subset of variables may be chosen using the option: **Each Y versus each X**

4.7.2   Matrix Created Selecting a Subset of Variables

Graph > Matrix Plot: Each Y versus each X

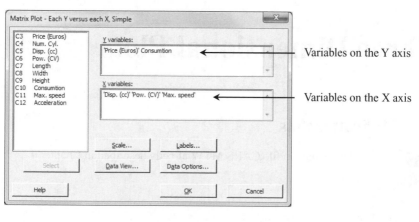

Variables on the Y axis

Variables on the X axis

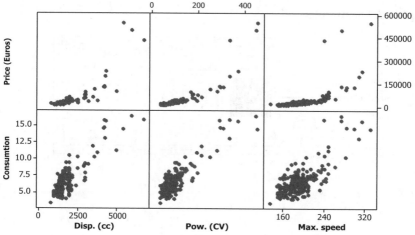

 In **Matrix Plots,** you can activate the **Brush** option to select a data point in one plot and Minitab automatically does the same across plots. Selecting data points in an array of plots is much easier than doing so having several graphs windows open at once.

5

Three Dimensional Plots

5.1 3D Scatterplots

 Consider the data in file CARS.MTW already described in Chapter 4.

Graph > 3D Scatterplot

3D scatterplot without
stratification ⟶

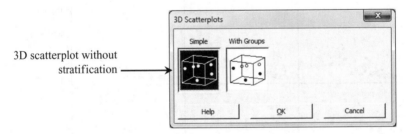

Industrial Statistics with Minitab, First Edition. Pere Grima Cintas, Lluís Marco-Almagro and Xavier Tort-Martorell Llabrés.
© 2012 John Wiley & Sons, Ltd. Published 2012 by John Wiley & Sons, Ltd.

Enter the price in **Z variable;** power in **Y variable**, and displacement in **X variable**. Using default options we obtain:

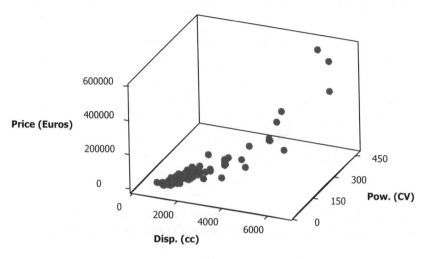

Notice that when the graph window opens, so also does the toolbar **3D Graph Tools**, which allows the graph to be acted on interactively.

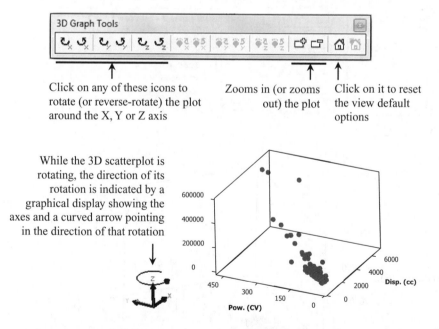

Click on any of these icons to rotate (or reverse-rotate) the plot around the X, Y or Z axis

Zooms in (or zooms out) the plot

Click on it to reset the view default options

While the 3D scatterplot is rotating, the direction of its rotation is indicated by a graphical display showing the axes and a curved arrow pointing in the direction of that rotation

The properties of the graph can be modified by double clicking on the item you want to change (points, axes, ...).

The Brush option can also be used in 3D scatterplots. To activate it, click on the indicated icon in the **Graph Editing** toolbar.

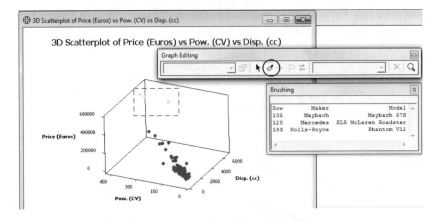

At times, it may be useful to draw lines joining each data point to its base. To do so, when constructing the graph, check the option **Project lines** in Data Display.

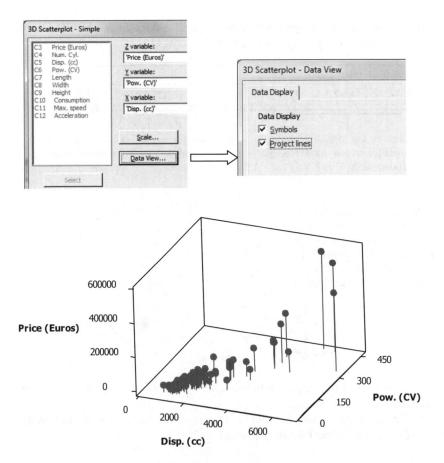

5.1.1   Stratification

Graph > 3D Scatterplot

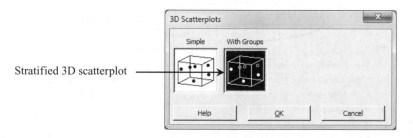

Stratified 3D scatterplot ⟶

The dialog box is similar to the previous case (without stratification), but it contains the new text box **Categorical variables for grouping (0–3)**. Entering there the variable number of cylinders ('Num.Cyl.'), we obtain:

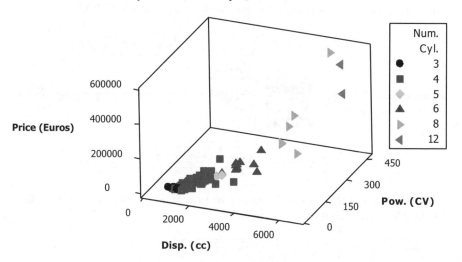

5.2   3D Surface Plots

Graph > 3D Surface Plot
Used to represent functions of the type z = f (x, y). To introduce the mesh of values (x, y), it is very useful to use the option: **Calc> Make Mesh Data**

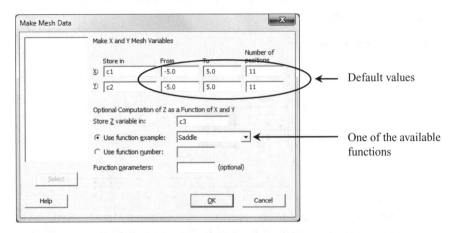

Default values

One of the available
functions

In addition to the built-in functions, you can also add others, the results of which
are a bit cumbersome because they require modifying an already built Minitab macro.

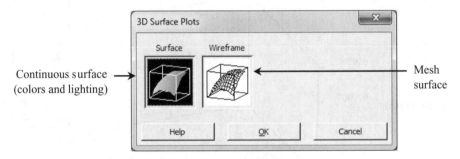

Continuous surface
(colors and lighting)

Mesh
surface

The dialog box that appears is exactly the same whether you choose the **Surface**
or the **Wireframe** option.

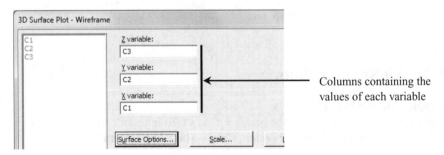

Columns containing the
values of each variable

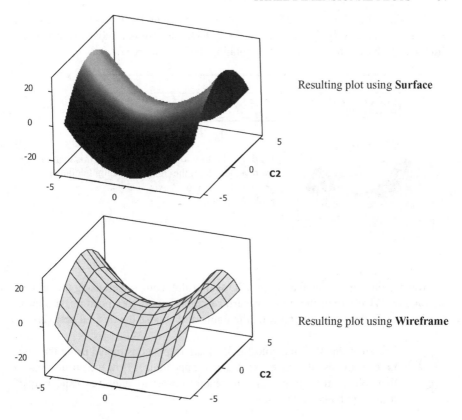

Resulting plot using **Surface**

Resulting plot using **Wireframe**

A **surface** graph can be transformed into a **wireframe** graph (and vice versa). To do so, right-click on the graph and choose the **Surface Type** option in **Edit Surface**

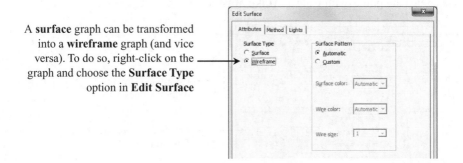

The axes of both types of 3D scatterplots, **Surface** and **Wireframe** graphs, can be rotated interactively. To do so, use the **3D Graph Tools** toolbar. Additionally, **Surface** graphs can be illuminated by rotating the light source.

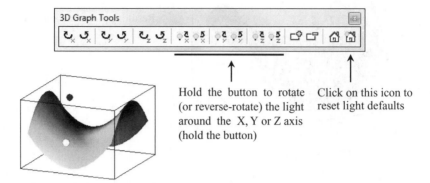

Hold the button to rotate (or reverse-rotate) the light around the X, Y or Z axis (hold the button)

Click on this icon to reset light defaults

You can customize any feature of a 3D scatterplot, constructed using the options **Surface** or **Wireframe**: the mesh or surface colors, spotlights position, the mesh density, etc. To do so, double-click on the figure and choose the desired options.

Presentations showing colorful 3D-scatterplots constructed using the **Surface** option usually have a spectacular appearance. However, bear in mind that **Wireframe** graphs are in general easier to interpret, especially if printed in black and white.

5.3   Contour Plots

Graph > Contour Plot
Generate data points with **Calc > Make Mesh Data:**

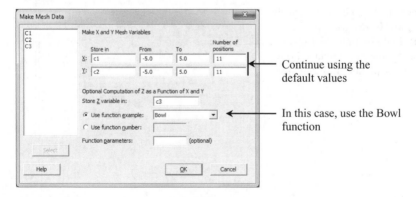

Continue using the default values

In this case, use the Bowl function

Graphing a mesh surface as follows.

Graph > 3D Surface Plot: Wireframe

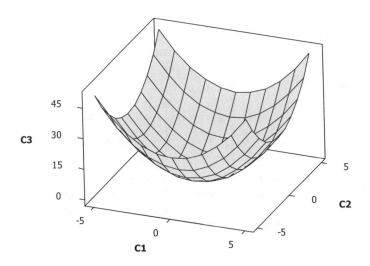

Graph > Contour Plot

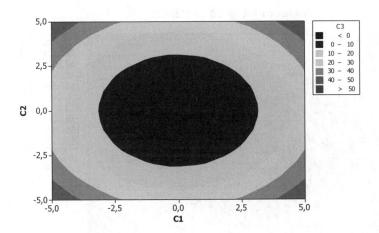

5.3.1 Contour Plot Options

Double-clicking over the contour plot, a dialog box pops up with options to change the filling color pattern and the number of levels.

Attributes tab   **Levels** tab

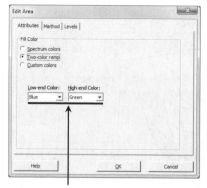
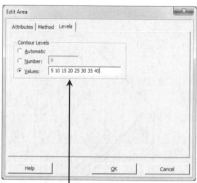

Choose the color pattern activating the
option **Two-color ramp** (by default
from blue to green). To select each
color individually, activate the **Custom
colors** option

Specify either the number of contour lines
(**Number**) or the values (**Values**) in which
to place such lines. The second option is
chosen in this case, using the shown values
with the format 5:40/5

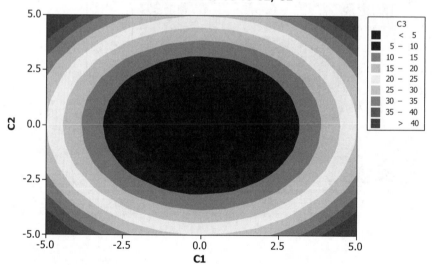

You can draw the contour lines without colors and indicating the value corresponding to each one of them. To do so, use the **Data View** options:

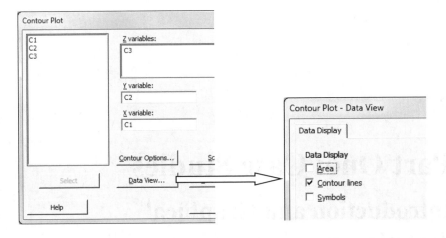

In **Contour Options**, you can indicate the values on which you want the contour lines.

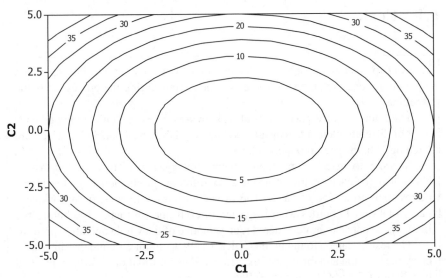

Contour Plot of C3 vs C2, C1

6

Part One: Case Studies
Introduction and Graphical Techniques

6.1   Cork

A cava (Spanish champagne) producing company decides to put into operation a plan to reduce the number of defects which occur during the process of bottle labelling.

The external appearance of a bottle is known as 'presentation'. It comprehends a set of different elements (capsule, collar, ring, label, back label, . . .) which are fixed up in high speed production lines.

In order to map out an improvement strategy, a plan to collect data was elaborated for each of the company's six different production lines (each line corresponds to a type of cava). The inspection lasted 15 days, during which a daily total of 100 bottles were inspected in each line (in total, 1500 bottles per line). These data can be considered representative of the whole operation process.

The results can be found in file CORK.MTW. Column C1 contains simultaneously the location and the type of the defects. This information is further specified in columns C2 and C3 using the following codes:

Industrial Statistics with Minitab, First Edition. Pere Grima Cintas, Lluís Marco-Almagro and Xavier Tort-Martorell Llabrés.
© 2012 John Wiley & Sons, Ltd. Published 2012 by John Wiley & Sons, Ltd.

Location (C2)		Type of defect (C3)	
1	Collar	1	Tear/bubble
2	Ring	2	Wrinkles
3	Label	3	Crooked
4	Back label	4	Alignment
5	Capsule	5	Height
6	Cap	6	Cap defect
7	Wire cage	7	Wire cage defect
		8	Open collar

Columns C4 to C9 correspond to each of the production lines under study. The purpose is to identify the production line and the type of problem with the highest number of defects and subsequently focus the attention on them.

The corresponding worksheet looks as follows:

↓	C1-T	C2	C3	C4	C5	C6	C7	C8	C9	C10
		Location	Defect	ALBA	TRADI	BLAU	1492	FIESTA	ROSADO	
1	Collar - tear/bubble	1	1	761	193	14	66	0	22	
2	Ring - tear/bubble	2	1	18	0	0	19	0	10	
3	Label - tear/bubble	3	1	69	92	14	53	48	16	
4	Back label - tear/bubble	4	1	19	25	3	19	26	4	
5	Collar - wrinkles	1	2	240	42	5	44	0	4	
6	Ring - wrinkles	2	2	34	2	0	24	0	11	
7	Label - wrinkles	3	2	66	8	1	9	13	5	
8	Back label - wrinkles	4	2	23	9	1	7	15	3	
9	Collar - crooked	1	3	76	34	9	22	0	7	
10	Ring - crooked	2	3	115	12	0	79	0	19	
11	Label - crooked	3	3	299	67	27	116	12	72	
12	Back label - crooked	4	3	30	3	11	54	49	3	
13	Capsule - alignment	5	4	242	32	0	85	65	15	
14	Label 2-4 - alignment	3	4	589	231	54	82	0	79	
15	Label > 4 - alignment	3	4	665	0	18	283	0	102	
16	Ring - alignment	2	4	631	135	0	237	0	45	
17	Back label 2-5 - alignment	4	4	401	21	5	158	18	85	
18	Back label > 5 - alignment	4	4	72	0	1	69	4	28	
19	Label - height	3	5	113	80	45	177	4	57	
20	Ring - height	2	5	332	58	0	209	0	42	
21	Collar - height	1	5	1	0	0	2	0	0	
22	Cap defect	6	6	0	0	0	0	0	0	
23	Wire cage	7	7	1	0	0	0	0	0	
24	Open collar	1	8	92	101	21	147	0	26	

To identify the production line with most defects produced, we sum up the values of columns C4 through C9. This can be done as follows.

Stat > Basic Statistics > Display Descriptive Statistics

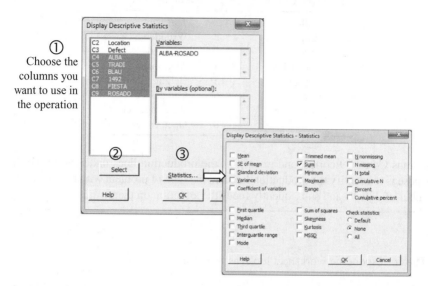

① Choose the columns you want to use in the operation

In the previous window, **Display Descriptive Statistics – Statistics**, the quickest way to only select the operation **Sum** is to disable all other statistics by activating the **None** option in **Check statistics** and, then, check **Sum**.

The computed sums appear in the Session window. They have to be copied and pasted into the worksheet.

Shortcut → Sum

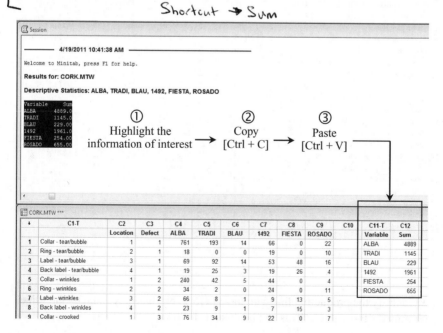

① Highlight the information of interest → ② Copy [Ctrl + C] → ③ Paste [Ctrl + V]

Once you have the data in this format, proceed to construct a Pareto diagram showing the contribution of each production line in the total number of defects, as follows.

Stat > Quality Tools > Pareto Chart

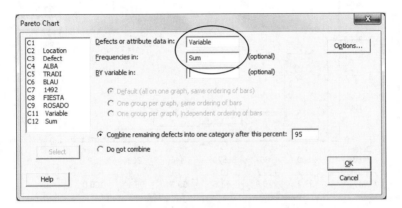

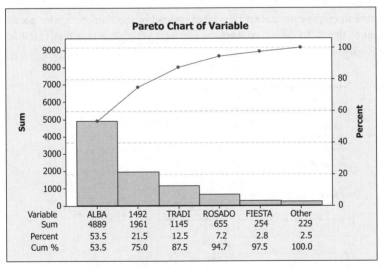

Clearly, the Alba line produces most defects (more than 50%), and both the Alba line and the 1492 line produce 75% of all defects. Focusing only on Alba line, it is also of interest to know the most frequent type of defect and location within this line. Hence, for the Alba line, we make a Pareto diagram of the type of defect:

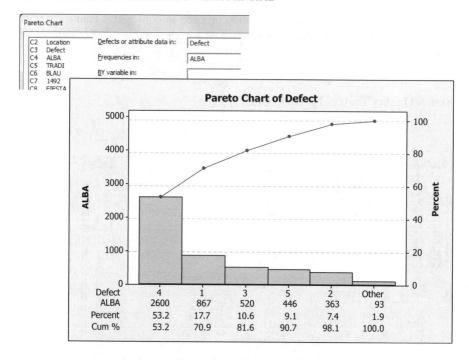

It is possible to display the defect type label instead of the numeric codes previously assigned to them. To do so, convert the numeric variable into a text variable, using the following.

✳Data > Code > Numeric to Text ✳

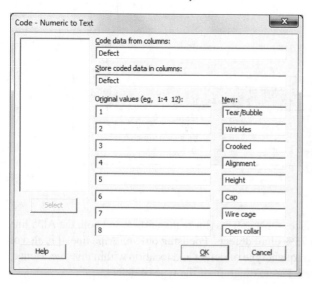

This modifies the contents of column C3, named 'Defect', and the new Pareto chart looks as follows:

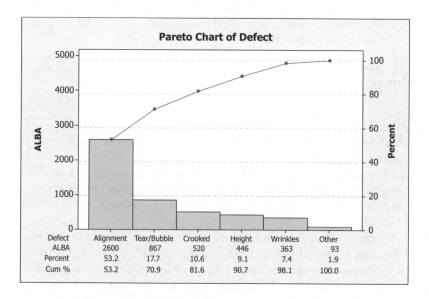

Pareto Chart of Defect

Defect	Alignment	Tear/Bubble	Crooked	Height	Wrinkles	Other
ALBA	2600	867	520	446	363	93
Percent	53.2	17.7	10.6	9.1	7.4	1.9
Cum %	53.2	70.9	81.6	90.7	98.1	100.0

As shown, the most frequent type of defect within the ALBA line is alignment. Let's see now which the most frequent location is. For this, as before, we start converting column C2 into a text variable (**Data > Code > Numeric to Text**). we then draw a Pareto diagram for the defect's location within the ALBA line. The resulting plot is the following:

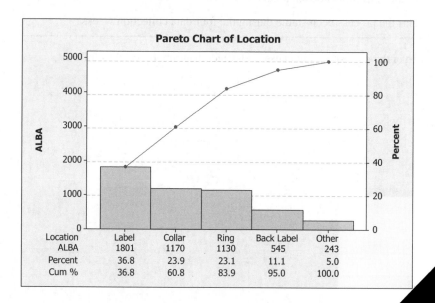

Pareto Chart of Location

Location	Label	Collar	Ring	Back Label	Other
ALBA	1801	1170	1130	545	243
Percent	36.8	23.9	23.1	11.1	5.0
Cum %	36.8	60.8	83.9	95.0	100.0

The location of the defects does not seem to be concentrated on a single element. A possible reason for this is that the most important type of defect is alignment, which affects several locations. The conclusion from the above analyses is that the most reasonable improvement approach would be to start focusing all efforts on the alignment problem in the ALBA line.

6.2 Copper

A manufacturer of copper tubes has detected, after collecting data and analyzing them with a Pareto diagram, that almost 70% of all process stops are due to breaks during the stretch of the tube.

After a brainstorming session attended by the heads of the stretching and casting sections, the three shift managers, and the person in charge of the laboratory, a list of all possible causes is compiled. These are shown in columns C1 to C6 of file COPPER.MTW.

Once the cause–effect diagram had been studied, it was decided to explore whether the alloy's contents of lead (Pb) or phosphorus (P) or the shift, which seemed to be the most probable causes, were really responsible for the breaks. In order to check that, data were collected throughout four weeks (60 shifts) on the following variables: the number of breaks, the contents (in ppm) of P and Pb in the alloy, and the shift within the breaks occurred.

Use Minitab to represent the cause–effect diagram and, analyzing the given data, to answer the following questions:

- Do the data confirm the suspicions that the content of P, the content of Pb or the shift influence the breaks?

- Has the process been stable during the four data collection weeks?

To construct the corresponding cause-and-effect diagram, proceed as follows.

Stat > Quality Tools > Cause-and-Effect

Columns that contain the primary causes   If appropriate, change default values

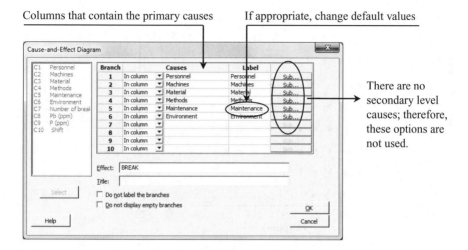

There are no secondary level causes; therefore, these options are not used.

Cause-and-Effect Diagram

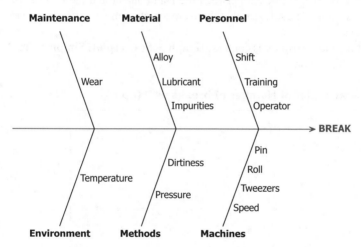

 The default font size has been enlarged (to do so, double-click on the group of letters you want to edit)

We can draw scatterplots in order to see if the contents of P, Pb, or the shift may have any influence on the number of breaks.

Relation with lead content: **Graph > Scatterplot: Simple > Y:** 'Number of breaks'; **X:** Pb (ppm).

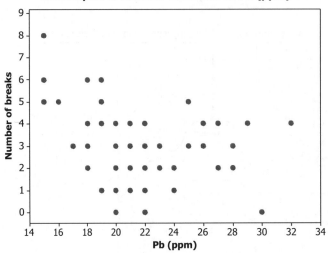

No relation seems to exist between the lead content and the number of breaks. Nonetheless, it may be relevant that there are a high number of breaks in case of low lead content.

Relation with phosphorus content: **Graph** > **Scatterplot: Simple** > **Y:** 'Number of breaks'; **X:** P (ppm).

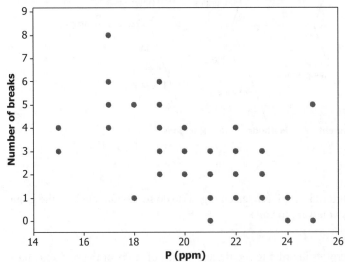

Since both variables are discrete, several points overlap and, hence, the density of points in each zone is not well represented. In order to avoid this problem, one can double-click on any point and then, in the **Edit symbols** dialog box, choose tab **Jitter** and check option **Add jitter to direction** (without changing any other default value). This implies a loss of precision for the sake of appreciating the density in each situation.

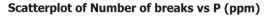

Scatterplot of Number of breaks vs P (ppm)

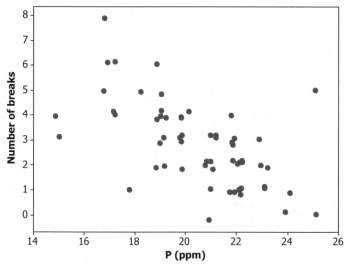

 In these graphs, the proportion between height and width has been changed (right-click on the exterior frame of the diagram: **Edit Figure Region** > **Graph Size**). The values of the horizontal scale have also been changed (double-click on any value of the scale).

There seems to be a correlation – not necessarily a cause–effect relation – between the number of breaks and the phosphorus content: the higher the content, the lower the number of breaks. Therefore, if increasing the phosphorus content is not contraindicated to other tube properties, it could be increased or left around 25 ppm to see whether a lower number of breaks is obtained.

The relation between the shift and the number of breaks will be analyzed by means of a type of graph we have not presented so far, which is very similar to a scatterplot but with a categorical variable in the abscissa:

Relation with the shift: **Graph** > **Individual Value Plot**

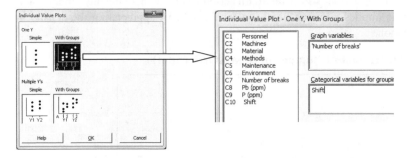

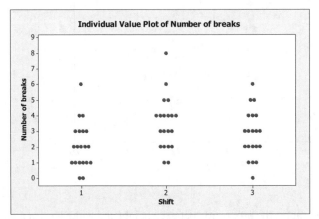

Even though one day 8 tubes broke in shift 2, the highest value among all, there is no apparent relation between shift and number of breaks.

To study whether the process remained stable throughout the four weeks, we can use a time series plot: **Graph** > **Time Series Plot: Simple**.

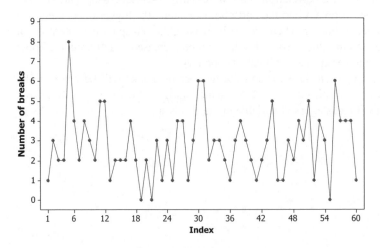

Since neither a trend nor a level change can be observed, it seems the process remained stable. Nonetheless, it is convenient to study the causes for which one day 8 breaks occurred.

It is possible to use different symbols for each shift. For this, double-click on any point, choose the **Groups** tab, **Categorical variables for grouping:** Shift. The result is the following graph, within which the symbol size was increased in the tab **Attributes (Symbols, Custom, Size:** 1.5).

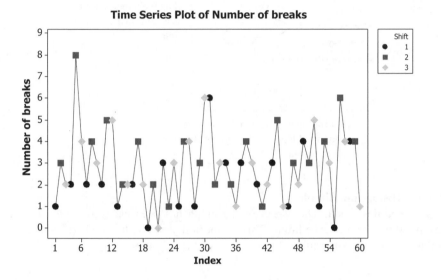

Time Series Plot of Number of breaks

6.3   Bread

The owner of a bakery, worried about the excessive variability of his products' weight, decided to carry out a study in order to analyze the distribution of the weight of one of his breads. The bread is elaborated by two operators (A and B) using two different machines (1 and 2). Both operators do not work simultaneously: the days A works, B does not and vice versa. For the study, daily samples of four pieces of bread of each machine were taken throughout a period of 20 days. The data obtained are included in file BREAD.MTW.

The nominal weight of the pieces of bread is 210 grams, but a variation of ± 10 grams is accepted. Which conclusions can be drawn from these data? Which recommendations would you give the owner of the bakery?

First of all, make a histogram of all data. For this, we have to pile all data in a single column.

Data > Stack > Columns

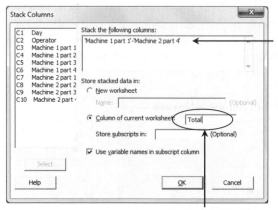

To automatically place all columns there, click on the name of the first column in the list on the left, then drag the mouse until the last, and finally click on **Select.**

Put the stacked data in the first empty column of the current worksheet and name it Total.

Graph > Histogram: Simple

Graph variables: Total. The aspect of the graph has been changed adding two lines that show the tolerance limits. This has been done using the toolbar: **Tools > Toolbars > Graph Annotation Tools**   *⚹ Click , Hold, Drag ⚹*

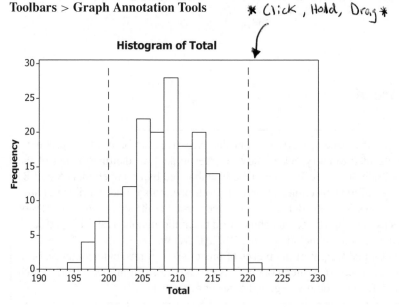

The distribution is left-skewed, with a certain percentage out of tolerances. Let us have a closer look stratifying by machine.

Put the values of Machine 1 in one column (**Data > Stack > Columns** for columns C3 to C6) and the ones of Machine 2 (C7 to C10) in another one. To make the histograms with an identical format, we can copy the one of the previous graph. With the histogram of all data as an active window, do the following.

Editor > Make Similar Graph

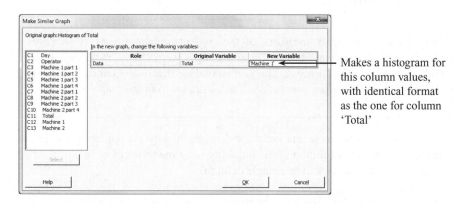

Makes a histogram for this column values, with identical format as the one for column 'Total'

Copying the format of the histogram for the column 'Total', also the lines which mark the specifications are copied. Since the vertical scale is shorter for each machine, these specification lines overhang the graph. The best is to delete them and draw them again.

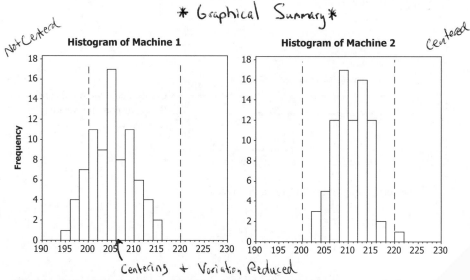

* Graphical Summary *

Not Centered

Centered

Centering + Variation Reduced

Machine 2 is centred whereas Machine 1 is not. Hence, the recommendation to bakery owner is to centre Machine 1; this would resolve a great part of the pr

To further improve, he should try to reduce the variability with which both machines produce.

Histograms could also be stratified by operator, or even by operator and machine, but the additional plots do not provide any relevant information.

6.4   Humidity

> During one week, daily measures of the content of humidity of 20 packages of a certain product were taken. These packages were chosen randomly at the end of the packaging process. The corresponding data is contained in the file HUMIDITY.MTW. What conclusions can be drawn from the graphical analyses of these data?

There are several options to analyze these data graphically. An appropriate graph to represent the weekly evolution of the humidity is a time series plot. To construct it, the data must first be stacked in a single column.

 Rarely is the available data already arranged in a convenient way for the analysis.

To pile up the data, proceed as usual: **Data > Stack > Columns**

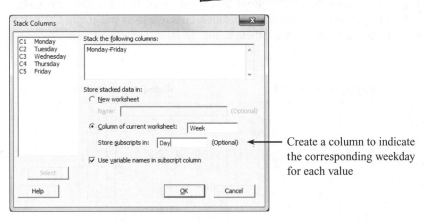

Create a column to indicate the corresponding weekday for each value

Time series diagram of the variable 'Week'.

Graph > Time Series Plot: Simple. Series: Week

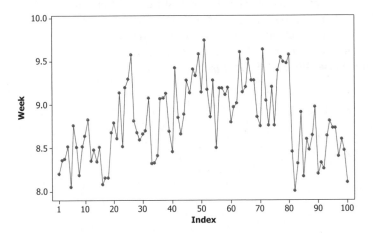

The <u>shown process is not stable</u>: there is first an increasing trend followed by a sudden jump downwards. To distinguish the weekdays, different line types and colors can be used. To do so, double-click on the line. A dialog box that allows you to edit the lines appears:

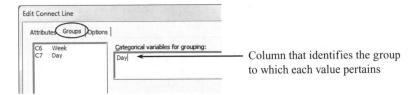

Column that identifies the group to which each value pertains

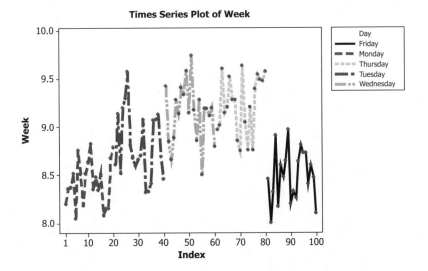

Times Series Plot of Week

Notice that, by default, the values of the legend (in our case, the weekdays) appear in alphabetic order. This order may be changed by first right-clicking on any cell of the corresponding column (in our case, column 'Day'), then choosing: **Editor** > **Column** > **Value Order** and finally introducing the desired order.

It is also possible to directly draw the stratified diagram using:

Graph > **Time Series Plot: With Groups**

Part Two

HYPOTHESIS TESTING. COMPARISON OF TREATMENTS

Part 2 begins with two short chapters which are somewhat unique because they are not directly devoted to data analysis. Chapter 7 is about generating random numbers, which allows the resolution of problems by simulation (in that chapter we offer a simple example), and testing what would happen if data had some regularity as indicated by a specific distribution. The chapter also explains how to enter data in a patterned way, something useful when we want to identify the origin of each observation. Chapter 8 is about computing probabilities with Minitab. It finishes with an application: calculating the sigmas of a process (the famous 3.4 ppm of a Six Sigma process appears here).

The core of Part 2 is dedicated to hypothesis testing, the name given to the reasoning procedure that we follow when performing a statistical test. The skeleton of this frequently-used reasoning procedure is the following:

1. State the null hypothesis and the alternative hypothesis. The null hypothesis is considered to be true unless the data (an objective representation of reality) are in contradiction with it. We do not prove that the null hypothesis is true; the test is designed to see if we have enough evidence to reject the null hypothesis, thus assuming then the certainty of the alternative hypothesis. Imagine we have data that we think comes from a Normal distribution, but we want to be sure that these data are not in contradiction with our hypothesis of normality (null hypothesis): conducting a normality test gives the answer.

2. From the available data, a value is computed that summarizes the discrepancy between the results obtained and the null hypothesis. If the discrepancy is very large, we reject the null hypothesis and we assume the correct one is the alternative. This measure is called the test statistic. We know that the values from a normal distribution represented on a normal probability plot form a

straight line. But when we have a sample, we should not expect a perfectly straight line (in the same way that we will not have a histogram with a perfect bell shape). The distance between our data points and the theoretical straight line is the test statistic. How to measure this distance depends on the type of test.

3. The distribution to test the statistic (reference distribution) is known when the null hypothesis is true. Some distributions are widely used for these purposes, as the normal, Student's t distribution or Snedecor's F distribution. In the case of normality tests, some other specific distributions are used. The key point is confronting the test statistic with the reference distribution. For instance, in normality tests we check if the discrepancy between data points and the straight theoretical line is reasonably small in case the data really come from a normal distribution.

4. Of course, it is not enough to determine if the discrepancy is big or small. We must quantify the discrepancy; this is done with the p-value. The p-value is a key number in statistical tests and gives the probability of having a discrepancy between our data and the null hypothesis as large as or larger than the one observed. If this probability is large, say 30%, we cannot say that our data are inconsistent with the null hypothesis. But if that probability is small, say 1 per 100000, the most reasonable choice is rejecting the null hypothesis. Where is the boundary between rejecting and not rejecting? Naturally it is an arbitrary value, but it is usually located at the 5% level.

All statistical tests follow this skeleton of reasoning. When comparing two means, the formula for calculating the test statistic depends on whether we have independent or paired samples, but in both cases the reference distribution is a Student's t distribution (degrees of freedom do depend on how we have the data). When comparing variances, the most common test statistic is the ratio between both variances, and the Snedecor's F-distribution is used as reference. Minitab also gives Levene's test result, which is useful when data is not normal. Comparison of proportions is usually done using an approximation to the normal distribution. This approach works well when $np>5$ and $np(1-p)>5$, where n is the sample size and p is the probability of success (following the nomenclature used with the binomial distribution, the appropriate one for this kind of variables). Minitab also presents the results of the 'Fisher exact test' that does not make this approximation.

The analysis of variance is a technique that deserves a separate chapter. Its name is misleading because its aim is not comparing variances but means. If we have two samples with means 10 and 15, the first with values ranging between 9 and 11, and the second between 14 and 16, it is reasonable to think that both treatments give different response levels. But imagine that, in both cases, values range from 2 to 20: then the difference is not so clear. That is, we compare the variability between treatments' means with the variability within treatments to reach conclusions on the equality of means. The test statistic is a variance ratio (maybe several if you have more than one factor) and the reference distribution is the Snedecor's F-distribution.

7

Random Numbers and Numbers Following a Pattern

7.1 Introducing Values Following a Pattern

At times – for instance when working with stratified data – it is convenient to create auxiliary variables containing data following a specific pattern. Minitab provides an easy way to create such type of data columns.

 Example 7.1: Place in column C1 numeric values going from 1 to 3 so that each value appears twice and the whole sequence four times, that is: 1, 1, 2, 2, 3, 3, 1, 1, 2, 2, 3, 3, 1, 1, 2, 2, 3, 3, 1, 1, 2, 2, 3, 3.

Calc > Make Patterned Data > Simple Set of Numbers

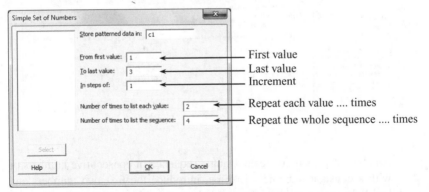

Alternatively, a more general way to create patterned data is the following.

Industrial Statistics with Minitab, First Edition. Pere Grima Cintas, Lluís Marco-Almagro and Xavier Tort-Martorell Llabrés.
© 2012 John Wiley & Sons, Ltd. Published 2012 by John Wiley & Sons, Ltd.

Calc > Make Patterned Data > Arbitrary Set of Numbers

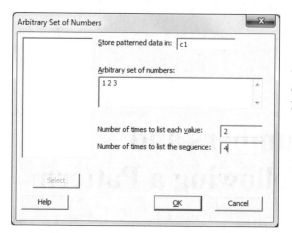

These values do not need to be equally spaced. For example, you could have the values 1, 2, 5.

In the dialog box that appears, first click on the **Help** button and then click on the hypertext '**arbitrary set of numbers**' to display a list of different ways of introducing the set of values that conform the sequence.

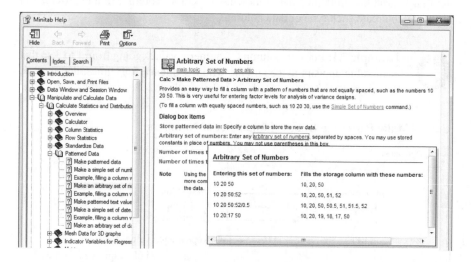

An alternative way to create a single sequence of consecutive numbers or with a constant increment is: First, introduce the first two numbers and highlight them with the mouse (click on the first number, then drag and hold the mouse). Second, place the pointer on the lower right corner of the highlighted area and finally, click and drag down the mouse (as in Excel).

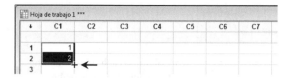

7.2   Sampling Random Data from a Column

Using any of the aforementioned procedures, introduce the numeric values 1 to 49 in column C1. To obtain a random sample of size six from these values, do the following.

Calc > Random Data > Sample From Columns
Fill the dialog box that appears, as indicated below:

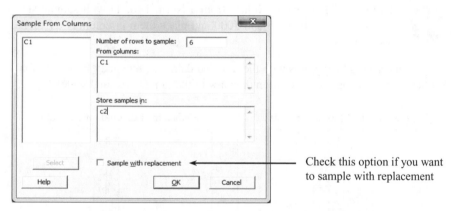

Check this option if you want to sample with replacement

This option could be used in simulation programs. A way to perform a simulation is using a Minitab macro, which allows the repetition of commands. A search on the Minitab help with the term 'macro' gives information about the different types of macros available.

7.3   Random Number Generation

Calc > Random Data > [Probability distribution of interest]
If we want to generate 1000 values from a normal distribution with $\mu = 500$ and $\sigma = 5$, and place them in columns C1 to C5, go to: **Calc > Random Data > Normal**:

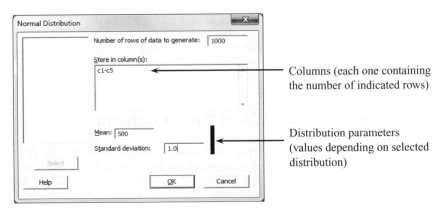

Columns (each one containing the number of indicated rows)

Distribution parameters (values depending on selected distribution)

 Example 7.2: Generate values from a Normal (0; 1) with sample sizes 50, 100, 250, 500, 1000 and 5000 and place them in columns C1 to C6, respectively.

When drawing histograms with the sampled data sets, they may look as follows (notice that all histograms have been drawn with the same horizontal scale):

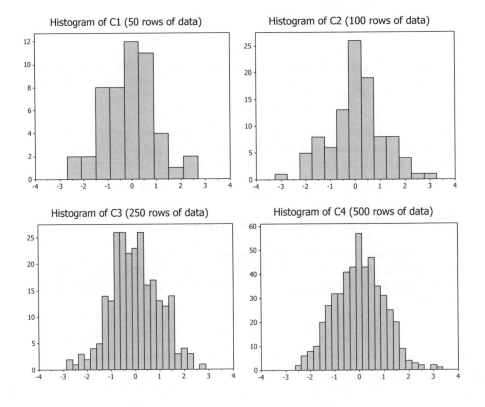

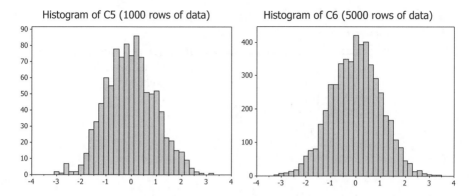

Histogram of C5 (1000 rows of data)   Histogram of C6 (5000 rows of data)

   To delete all plots at once, go to: **Window > Close All Graphs (OK, No to All)**.

7.4   Example: Solving a Problem Using Random Numbers

A common operation in the assembling process of a certain mechanism consists of introducing the ending part of a shaft into a given hole; the end shaft must stay inside the hole. If the hole diameter X is distributed according to a normal N(3,00; 0,03) and the shaft diameter Y according to a N(2,98; 0,04), what percentage of these operations will not be carried out because the shaft is thicker than the hole?

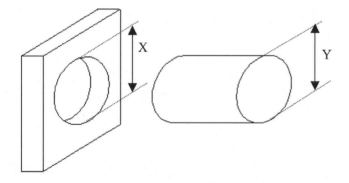

Generate 10000 axis-values from a normal N (3.00; 0.03) and place them in column C1. Likewise, generate 10000 hole-values from a normal N (2.98, 0.04) and store them in C2, so that we end with 10000 couples of axis-hole. Those couples in which C1 values are less than their respective C2 values cannot be mounted.

To find out in how many cases the values of C1 are less than their respective row values in C2, use the calculator logical functions: **Calc > Calculator**

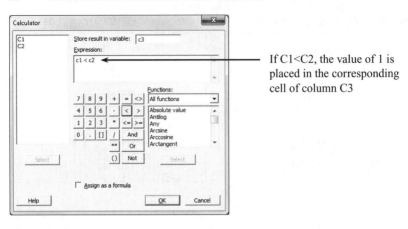

If C1<C2, the value of 1 is placed in the corresponding cell of column C3

Then, just add the values in C3 (**Calc > Column Statistics**) and divide by 10000. The result, computed analytically, is 34.46%.

8

Computing Probabilities

8.1 Probability Distributions

Calc > Probability Distributions

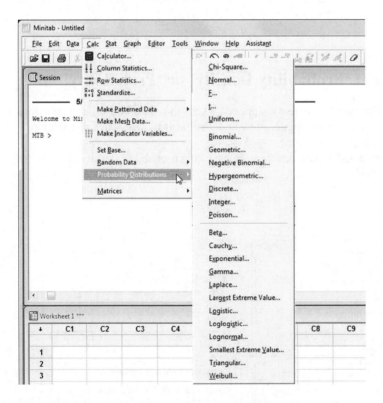

Industrial Statistics with Minitab, First Edition. Pere Grima Cintas, Lluís Marco-Almagro and Xavier Tort-Martorell Llabrés.
© 2012 John Wiley & Sons, Ltd. Published 2012 by John Wiley & Sons, Ltd.

Among other possibilities, this option allows you to determine the probabilities associated with the available distributions.

For example, to select the Normal distribution, do the following:

Calc > Probability Distributions > Normal

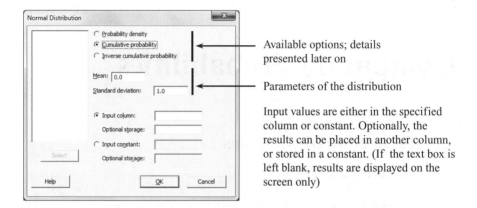

Available options; details presented later on

Parameters of the distribution

Input values are either in the specified column or constant. Optionally, the results can be placed in another column, or stored in a constant. (If the text box is left blank, results are displayed on the screen only)

8.2   Option 'Probability Density' or 'Probability'

The option **Probability density** appears when the distribution is continuous (Normal, Student's t, Chi-square, etc.) and **Probability** when the distribution is discrete (binomial, Poisson, etc.).

Probability density provides the ordinate of the probability density function for the given input value. That is:

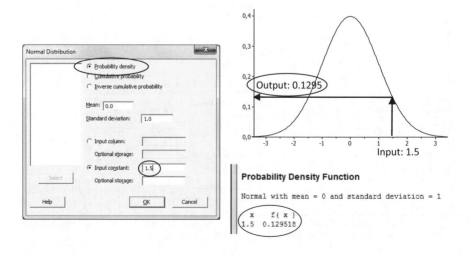

Probability is the option that appears when the distribution is discrete, and provides the probability associated with the input value.

8.3   Option 'Cumulative Probability'

Provides the probability of having a value less than or equal to the input. Example:

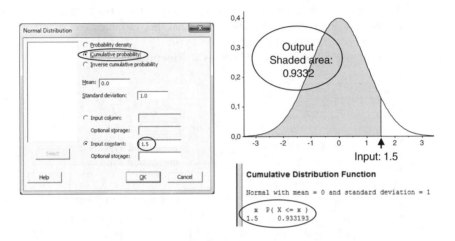

8.4   Option 'Inverse Cumulative Probability'

Contrary to the previous option, this provides the value of the variable associated with a specific cumulative probability:

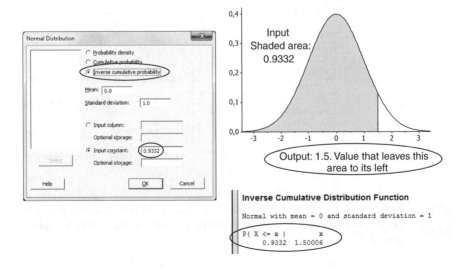

Be aware not to use the option **Probability density** to compute probabilities in a continuous distribution.

Example 8.1: A machine produces 5% of defective parts. If packed in boxes of 100 units, calculate:

a) The probability that a box has exactly three defective units.

The number of defective units in a box is a random variable that follows a binomial distribution. The probability that this variable takes a certain value (3) is of interest now:

Calc > Probability Distributions > Binomial

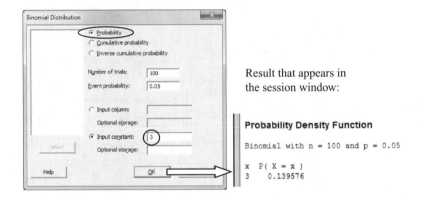

Result that appears in the session window:

Probability Density Function

Binomial with n = 100 and p = 0.05

```
x   P( X = x )
3    0.139576
```

b) The probability that a box has less than three defective units.

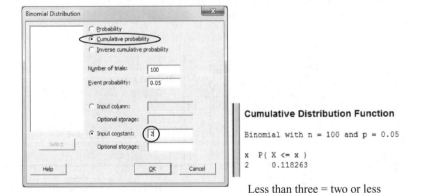

Cumulative Distribution Function

Binomial with n = 100 and p = 0.05

```
x   P( X <= x )
2     0.118263
```

Less than three = two or less

Example 8.2: A packaging process specializes in filling containers. The weight of containers is normally distributed with a mean of 975 g and a standard deviation of 10 g. Calculate:

a) The probability that a container has a weight less than 960 g.

Calc > Probability Distributions > Normal

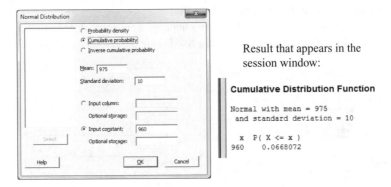

Result that appears in the session window:

Cumulative Distribution Function

```
Normal with mean = 975
  and standard deviation = 10

    x    P( X <= x )
  960    0.0668072
```

b) The probability that a container has a weight between 960 and 1040 g.

Similarly to item a), by replacing 960 by 1040, you can compute the probability of having a container with weight less than 1040 g. The obtained probability value is 1.00000; this means that practically all containers have a weight below 1040 g:

Cumulative Distribution Function

```
Normal with mean = 975 and standard deviation = 10

    x    P( X <= x )
 1040      1.00000
```

The probability that a container has a weight between 960 and 1040 g. is given by: $1.00 - 0.07 = 0.93$

c) What weight will be exceeded by 5% of the containers? In other words, you should determine the weight that leaves below 95% of the containers. That is:

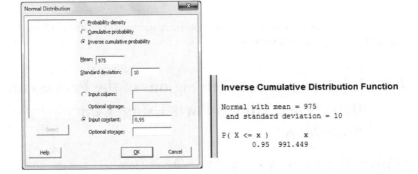

Inverse Cumulative Distribution Function

```
Normal with mean = 975
  and standard deviation = 10

P( X <= x )        x
   0.95      991.449
```

8.5   Viewing the Shape of the Distributions

The shape of probability distributions can easily be represented using the option:
Graph > Probability Distribution Plot

For example, to illustrate how the shape of a Weibull distribution varies as a function of the shape parameter, with a fixed scale parameter equal to 1, do:

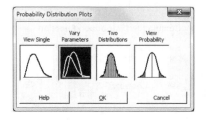
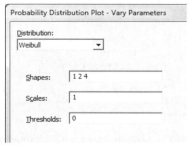

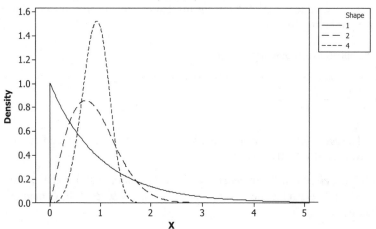

Distribution Plot
Weibull, Scale=1, Thresh=0

There are many options available that the reader could easily explore, especially useful for pedagogical purposes.

8.6   Equivalence between Sigmas of the Process and Defects per Million Parts Using *'Cumulative Probability'*

A process is said to be 'Six Sigma' when the distance between the nominal value and the tolerance limits of the produced output is equal to six times the standard deviation with which the output is produced. Likewise, a process is 'Five Sigma' if the tolerance limits are at five standard deviations, etc.

A Six Sigma process would look as follows:

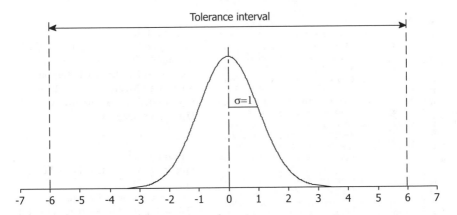

Knowing that the process will not always remain centred on its nominal value, it is assumed that the process is decentred 1.5 standard deviations to compute the proportion of defects that are produced. That is:

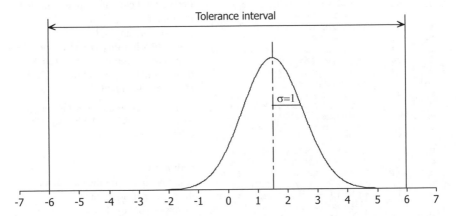

To compute the proportion of defects in relation to the sigmas of the process, set $\sigma = 1$ and vary the tolerance values. Tolerances ± 6 correspond to a 6σ process, ± 5 to a 5σ process, etc.

↓	C1	C2	C3	C4	C5	C6
1	6,0	-6,0	0,9999966	0,0000034	0,0000000	0,0000034
2	5,9	-5,9	0,9999946	0,0000054	0,0000000	0,0000054
3	5,8	-5,8	0,9999915	0,0000085	0,0000000	0,0000085
4	5,7	-5,7	0,9999867	0,0000133	0,0000000	0,0000133
5	5,6	-5,6	0,9999793	0,0000207	0,0000000	0,0000207
6	5,5	-5,5	0,9999683	0,0000317	0,0000000	0,0000317
7	5,4	-5,4	0,9999519	0,0000481	0,0000000	0,0000481
8	5,3	-5,3	0,9999277	0,0000723	0,0000000	0,0000723
9	5,2	-5,2	0,9998922	0,0001078	0,0000000	0,0001078
10	5,1	-5,1	0,9998409	0,0001591	0,0000000	0,0001591
11	5,0	-5,0	0,9997674	0,0002326	0,0000000	0,0002326
12	4,9	-4,9	0,9996631	0,0003369	0,0000000	0,0003369
13	4,8	-4,8	0,9995166	0,0004834	0,0000000	0,0004834
14	4,7	-4,7	0,9993129	0,0006871	0,0000000	0,0006871
15	4,6	-4,6	0,9990324	0,0009676	0,0000000	0,0009676
16	4,5	-4,5	0,9986501	0,0013499	0,0000000	0,0013499
17	4,4	-4,4	0,9981342	0,0018658	0,0000000	0,0018658
18	4,3	-4,3	0,9974449	0,0025551	0,0000000	0,0025551
19	4,2	-4,2	0,9965330	0,0034670	0,0000000	0,0034670
20	4,1	-4,1	0,9953388	0,0046612	0,0000000	0,0046612
21	4,0	-4,0	0,9937903	0,0062097	0,0000000	0,0062097
22	3,9	-3,9	0,9918025	0,0081975	0,0000000	0,0081976
23	3,8	-3,8	0,9892759	0,0107241	0,0000001	0,0107242
24	3,7	-3,7	0,9860966	0,0139034	0,0000001	0,0139035
25	3,6	-3,6	0,9821356	0,0178644	0,0000002	0,0178646
26	3,5	-3,5	0,9772499	0,0227501	0,0000003	0,0227504
27	3,4	-3,4	0,9712834	0,0287166	0,0000005	0,0287170
28	3,3	-3,3	0,9640637	0,0359303	0,0000008	0,0359311
29	3,2	-3,2	0,9554345	0,0445655	0,0000013	0,0445668
30	3,1	-3,1	0,9452007	0,0547993	0,0000021	0,0548014
31	3,0	-3,0	0,9331928	0,0668072	0,0000034	0,0668106
32	2,9	-2,9	0,9192433	0,0807567	0,0000054	0,0807621
33	2,8	-2,8	0,9031995	0,0968005	0,0000085	0,0968090
34	2,7	-2,7	0,8849303	0,1150697	0,0000133	0,1150830
35	2,6	-2,6	0,8643339	0,1356661	0,0000207	0,1356867
36	2,5	-2,5	0,8413447	0,1586553	0,0000317	0,1586869
37	2,4	-2,4	0,8159399	0,1840601	0,0000481	0,1841082
38	2,3	-2,3	0,7881446	0,2118554	0,0000723	0,2119277
39	2,2	-2,2	0,7580363	0,2419637	0,0001078	0,2420715
40	2,1	-2,1	0,7257469	0,2742531	0,0001591	0,2744122
41	2,0	-2,0	0,6914625	0,3085375	0,0002326	0,3087702
42	1,9	-1,9	0,6554217	0,3445783	0,0003369	0,3449152
43	1,8	-1,8	0,6179114	0,3820886	0,0004834	0,3825720
44	1,7	-1,7	0,5792597	0,4207403	0,0006871	0,4214274
45	1,6	-1,6	0,5398278	0,4601722	0,0009676	0,4611398
46	1,5	-1,5	0,5000000	0,5000000	0,0013499	0,5013499
47	1,4	-1,4	0,4601722	0,5398278	0,0018658	0,5416937
48	1,3	-1,3	0,4207403	0,5792597	0,0025551	0,5818148
49	1,2	-1,2	0,3820886	0,6179114	0,0034670	0,6213784
50	1,1	-1,1	0,3445783	0,6554217	0,0046612	0,6600829
51	1,0	-1,0	0,3085375	0,6914625	0,0062097	0,6976721

Columns C1 and C2:
Tolerance limits

Calc > Make Patterned Data

Column C3:
Proportion below the value indicated in column C1
Calc > Probability Distributions > Normal: Mean: 1.5. Standard deviation: 1. Input column: C1. Optional storage: C3

Column C4:
Proportion above the value indicated in column C1
(outside tolerances; by excess)
Calc > Calculator. Store result in C4.
Expression: 1-C3

Column C5:
Proportion below the value indicated in column C2.
(outside tolerances; by defect)
Calc > Probability Distributions > Normal. Mean: 1,5. Standard deviation: 1. Input column: C2. Optional storage: C5

Note: Some tables do not include this part because its value is negligible when the number of sigmas is high

Column C6:
Add together (with the calculator) the two types of defects:
by excess and by defect.
C6 = C4 + C5

The default number of decimal points has been changed in columns C3 to C6: Highlight the columns. **Editor > Format Column > Numeric ...**

Column C6 contains the proportion of defects. To convert to ppm multiply these values by 1000000. For example, a process 3.5σ produces 22750 ppm.

9

Hypothesis Testing for Means and Proportions. Normality Test

9.1 Hypothesis Testing for One Mean

9.1.1 Case 1: Known Population Standard Deviation

Stat > Basic Statistics > 1-Sample Z

Example 9.1: A filling line of detergent packages must introduce 4 kg in each package. A sample of 20 units is taken and subsequently weighted (in grams) obtaining the following measures:

| 4035 | 3974 | 3949 | 4009 | 3969 | 3970 | 3955 | 4034 | 3969 | 3991 |
| 3928 | 4024 | 4017 | 3983 | 3979 | 3997 | 3984 | 3964 | 3995 | 3988 |

Historical data indicate a standard deviation of the weights of 25 g. Check if the process is not centred; that is, check if the mean of the packets' weights is different from the target value of 4 kg.

Enter the data in column C1 and call it 'Weights'. Then, go to **Stat > Basic Statistics > 1-Sample Z**:

Industrial Statistics with Minitab, First Edition. Pere Grima Cintas, Lluís Marco-Almagro and Xavier Tort-Martorell Llabrés.
© 2012 John Wiley & Sons, Ltd. Published 2012 by John Wiley & Sons, Ltd.

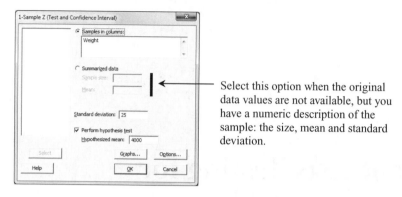

Select this option when the original data values are not available, but you have a numeric description of the sample: the size, mean and standard deviation.

With all default options, the output is the following:

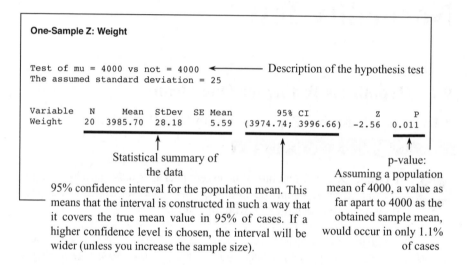

One-Sample Z: Weight

Test of mu = 4000 vs not = 4000 ◄─────── Description of the hypothesis test
The assumed standard deviation = 25

Variable	N	Mean	StDev	SE Mean	95% CI	Z	P
Weight	20	3985.70	28.18	5.59	(3974.74; 3996.66)	-2.56	0.011

Statistical summary of
the data

p-value:
Assuming a population
mean of 4000, a value as
far apart to 4000 as the
obtained sample mean,
would occur in only 1.1%
of cases

95% confidence interval for the population mean. This means that the interval is constructed in such a way that it covers the true mean value in 95% of cases. If a higher confidence level is chosen, the interval will be wider (unless you increase the sample size).

The options are:

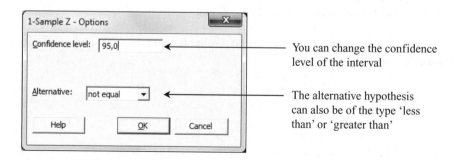

You can change the confidence level of the interval

The alternative hypothesis can also be of the type 'less than' or 'greater than'

Notice that the button **Graphs** allows you to select any of the following graphs:

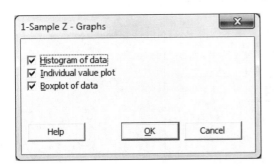

All these graphs show the sample mean ($\bar{x}$), the confidence interval (using the chosen confidence level) and the value of the null hypothesis.

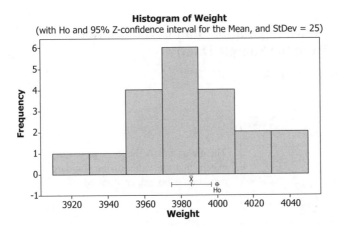

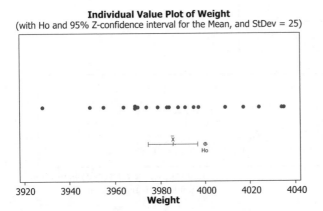

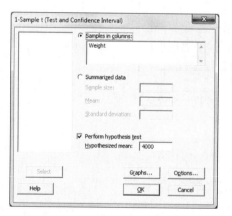

Boxplot of Weight
(with Ho and 95% Z-confidence interval for the Mean, and StDev = 25)

9.1.2 Case 2: Unknown Population Standard Deviation (Estimated from the Sample Data)

Stat > Basic Statistics > 1-Sample t

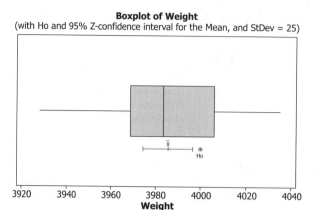

The dialog box is the same as before, except that the value of the standard deviation is no longer required

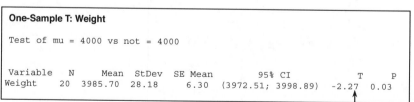

One-Sample T: Weight

```
Test of mu = 4000 vs not = 4000


Variable   N     Mean   StDev   SE Mean      95% CI          T     P
Weight     20   3985.70  28.18    6.30   (3972.51; 3998.89)  -2.27  0.03
```

In this case, the reference distribution is Student's t-distribution

9.2 Hypothesis Testing and Confidence Interval for a Proportion

Stat > Basic Statistics > 1 Proportion

Example 9.2: A certain consumer electronics product offers a new feature that increases its price. Nonetheless, it seems that only very few people use that feature. To confirm that, a survey is carried out among 200 users and it is found that only 17 clients use that feature. We are interested in:

1. Calculating a 95% confidence interval for the proportion of users that use the feature in question.

2. Do the data confirm the suspicion that less than 10% of users are making use of the new feature?

In this case, you do not need to enter the data in the worksheet. Instead, go directly to **Stat > Basic Statistics > 1 Proportion:**

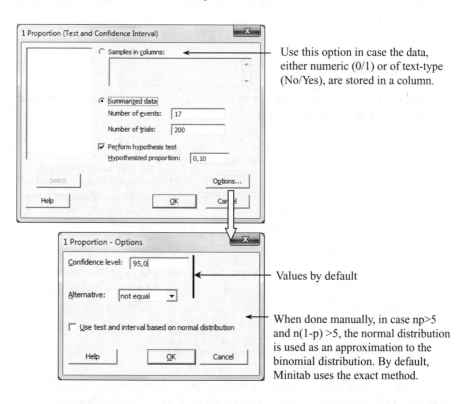

Use this option in case the data, either numeric (0/1) or of text-type (No/Yes), are stored in a column.

Values by default

When done manually, in case np>5 and n(1-p) >5, the normal distribution is used as an approximation to the binomial distribution. By default, Minitab uses the exact method.

The result is:

```
Test of p = 0.1 vs p not = 0.1
                                                  Exact
Sample  X    N  Sample p         95% CI          P-Value
1       17  200  0.085000  (0.050296; 0.132605)   0.487
```

Notice that the problem formulation guides us to choose the option 'less than' as alternative hypothesis, obtaining:

```
Test of p = 0.1 vs p < 0,1
                               95% Upper    Exact
Sample  X    N  Sample p         Bound    P-Value
1       17  200  0.085000      0.124771    0.285
```

Given the obtained p-value of 0.285, clearly larger than 0.05, the survey data do not allow us to state that the percentage of users that use the new feature in question is less than 10%.

9.3   Normality Test

To apply this test to the data of the detergent's weights presented before, go to: **Stat > Basic Statistics > Normality Test**

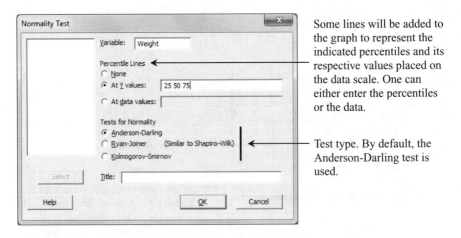

Some lines will be added to the graph to represent the indicated percentiles and its respective values placed on the data scale. One can either enter the percentiles or the data.

Test type. By default, the Anderson-Darling test is used.

If the data come from a normal distribution, the values of their cumulative proportions will form approximately a straight line on a normal probability plot.

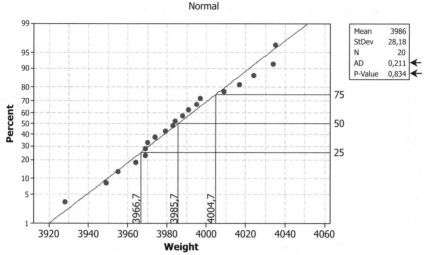

Probability Plot of Weight
Normal

AD: The Anderson-Darling statistic (AD) is computed as a function of the distances of the points and the straight line. The larger this distance, the greater is the value of AD.

p-value: If the data follow a normal distribution, a discrepancy between the straight line and the points (measured by the AD) as large as the observed or greater, occurs in 83.4% of cases.

An alternative way to carry out the normality test is the following.

Graph > Probability Plot: Single

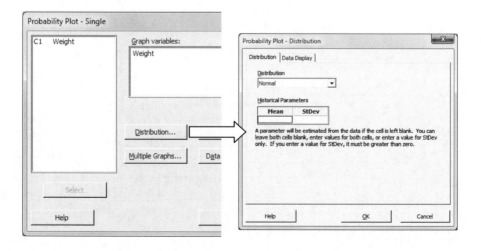

Notice that it is possible to fit other distributions. According to the chosen distribution, its parameters may be entered; however, this is optional as by default they are automatically estimated from the data.

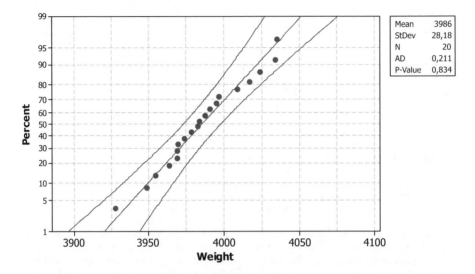

If the points lie within the confidence bands, it can be assumed that the values come from the distribution under study using a confidence level of 95%. This level can be changed in tab **Data Display**, the option which is found next to the **Distribution** tab.

 If only few data are available, which is equivalent to having little information, the hypothesis that the data come from the distribution under study is almost never rejected; independently of the used distribution.

10

Comparison of Two Means, Two Variances or Two Proportions

10.1 Comparison of Two Means

10.1.1 Case 1: Independent Samples

Stat > Basic Statistics > 2-Sample t

Example 10.1: Prat *et al.* (2004), in *Métodos Estadísticos. Control y mejora de la calidad* (Ediciones UPC, 2004), present a case study dealing with the comparison of two products (A and B) that are used in a leather tanning process. Ten pieces of leather are tanned using product A, another 10 with B. Once all 20 pieces are tanned the tensile strength of each piece is measured, obtaining the following data (in units of the measuring instrument):

Tanning using A: 24.3 25.6 26.7 22.7 24.8 23.8 25.9 26.4 25.8 25.4
Tanning using B: 24.4 21.5 25.1 22.8 25.2 23.5 22.2 23.5 23.3 24.7

Can we affirm that both products give different results with respect to the tensile strength?

Place the values of samples A and B in columns C1 and C2, respectively. Then, go to: **Stat > Basic Statistics > 2-Sample t**

Industrial Statistics with Minitab, First Edition. Pere Grima Cintas, Lluís Marco-Almagro and Xavier Tort-Martorell Llabrés.
© 2012 John Wiley & Sons, Ltd. Published 2012 by John Wiley & Sons, Ltd.

Depending on how the data is organized, choose one of the three following options:

- **Samples in one column:** Choose this option when all data are contained in a single column (**Samples**) and another column (**Subscripts**) is used to indicate the corresponding treatment (for that, numeric codes like 1 or 2 may be used).

- **Samples in different columns**: Use this option when the values of each sample are contained in different columns. In our case, this option is chosen since we have placed the values of samples A and B in columns C1 and C2, respectively.

- **Summarized data**: Select this option when the original data values are not available, but you have a numeric description of the samples: size, mean and standard deviation of each sample.

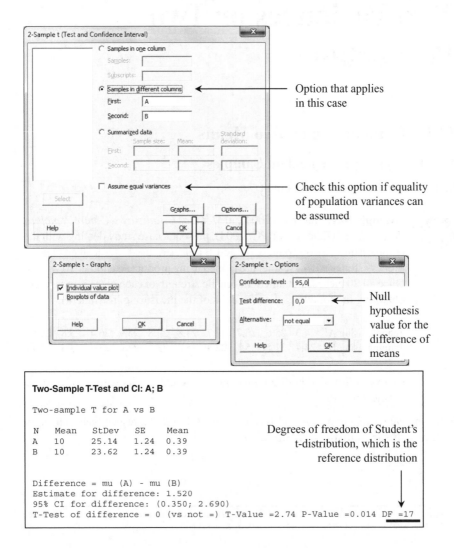

If help over a specific topic is needed, for example if you want to find out how the value of the degrees of freedom for the student's t-distribution is obtained (reference distribution with DF = 17), go to: **Help > Help > Index > 2-Sample t > see also > Methods and formulas > Test statistics**.

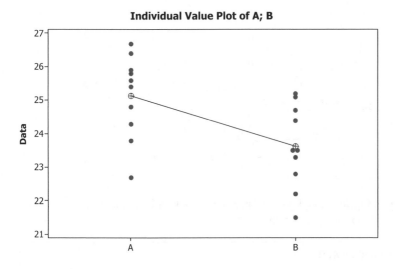

Recall that a visual inspection of the data is highly recommended. Indeed, a suitable graphical representation will allow, for example, the detection of anomalies.

10.1.2   Case 2: Paired Data

Stat > Basic Statistics > Paired t

Example 10.2: Prat *et al.* (2004), in *Métodos Estadísticos. Control y mejora de la calidad* (Ediciones UPC, 2004), consider a case study about the comparison of two superficial treatments for lenses with respect to the degree of deterioration. Ten subjects, who are wearing glasses, are selected for the study and received each a lens treated with treatment A and a lens treated with B (the position, either left or right, is randomly chosen). After a certain time, a measure of the degree of deterioration (scratches, wear of the superficial layer, . . .) of each lens is taken. The obtained data are:

Subject:	1	2	3	4	5	6	7	8	9	10
Treatment A:	6.7	5.0	3.6	6.2	5.9	4.0	5.2	4.5	4.4	4.1
Treatment B:	6.9	5.8	4.1	7.0	7.0	4.6	5.5	5.0	4.3	4.8

Can we affirm that the degree of deterioration depends on the received treatment?

Place the values of samples A and B in columns C1 and C2, respectively. Then, go to: **Stat > Basic Statistics > Paired t**.

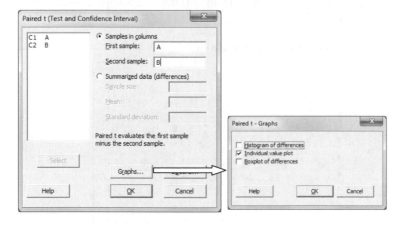

Clicking on **Options** displays the same options appeared in the **2-Sample t** case. Using all default options, we obtain:

```
Paired T-Test and CI: A; B

Paired T for A - B
                N    Mean   StDev   SE Mean
A              10   4.960   1.030     0.326
B              10   5.500   1.130     0.357
Difference     10  -0.540   0.344     0.109

95% CI for mean difference: (-0.786; -0.294)
T-Test of mean difference =0 (vs not=0): T-Value =-4.97   P-Value =0.001
```

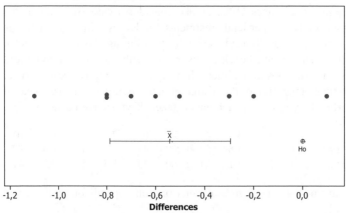

Individual Value Plot of Differences
(with Ho and 95% t-confidence interval for the mean)

The p-value (=0.001) clearly indicates rejection of the null hypothesis of equality of means; the default significance level (alpha = 0.05) is used. To illustrate this result, an individual value plot of the paired differences and a 95% confidence interval for the mean of the differences are displayed. Additionally, the null hypothesis test value (difference = 0) is indicated on the plot. Notice that the null hypothesis value is not included in the confidence interval, which confirms that in this case the null hypothesis is rejected with a p-value smaller than the specified level of significance (p-value = 0.001 < alpha = 0.05).

 It is essential to distinguish whether you are dealing with independent data (**2-Sample t** test would be the adequate option) or with paired data (**Paired t** option). This of course depends on the experimental design and consequently on the way the data is collected.

10.2   Comparison of Two Variances

Stat > Basic Statistics > 2 Variances
We apply this test to the tensile strength data in Example 10.1, although we already observed that the sample variances are practically equal and thus this suggests that the test of equality of variances will not be rejected.

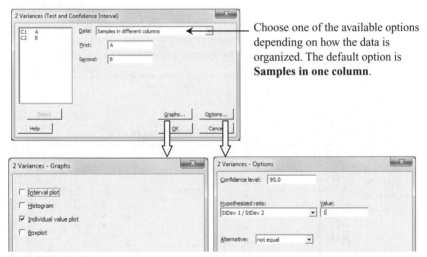

Choose one of the available options depending on how the data is organized. The default option is **Samples in one column**.

By default, all plots are chosen. The information you get is a bit redundant

You can choose to compare variances or standard deviations (the latter by default). In practice, it is not relevant the choice of one option over the other.

Value: Ratio of values that are contrasted, 1 (default value) means equality.

Test and CI for Two Variances: A; B

```
Method
Null hypothesis        Sigma(A) / Sigma(B) = 1                    Statement
Alternative hypothesis Sigma(A) / Sigma(B) not = 1
Significance level     Alpha = 0.05


Statistics

Variable   N   StDev   Variance
A         10   1.242   1.543                          Descriptive statistics
B         10   1.237   1.531

Ratio of standard deviations = 1.004
Ratio of variances = 1.008


95% Confidence Intervals
--                                              Confidence intervals:
                                CI for           If they include the value of one,
Distribution   CI for StDev    Variance          the null hypothesis of equality of
of Data           Ratio          Ratio          variances cannot be rejected (with
Normal         (0.500; 2.014)  (0.250; 4.058)        this level of confidence).
Continuous     (0.369; 2.243)  (0.136; 5.030)

Tests
                                           Test
Method                         DF1  DF2  Statistic  P-Value
F Test (normal)                  9    9    1.01      0.991
Levene's Test (any continuous)   1   18    0.00      0.955
                                                           Tests:
```

The F-test is the most commonly used, but it requires normality of data. The Levene test only requires data coming from a continuous distribution.

The graphical analysis corroborates that the variances cannot be considered different.

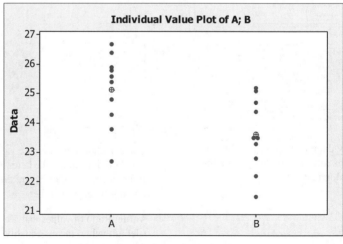

Individual Value Plot of A; B

10.3   Comparison of Two Proportions

Stat > Basic Statistics > 2 Proportions

Example 10.3: A survey has been carried out among 300 clients of a northern region and it turned out that 33 of them are unsatisfied with the company's service. In the southern region, 250 clients have been consulted and 22 of them expressed discontent. Given this information, can we affirm that the proportion of unsatisfied clients differs between both regions?

In this case, it is not necessary to enter values in the worksheet, since they are entered directly in the dialog box.

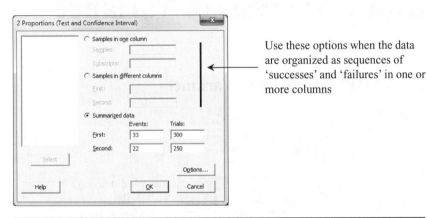

Use these options when the data are organized as sequences of 'successes' and 'failures' in one or more columns

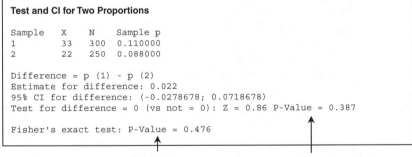

```
Test and CI for Two Proportions

Sample   X    N    Sample p
1        33   300  0.110000
2        22   250  0.088000

Difference = p (1) - p (2)
Estimate for difference: 0.022
95% CI for difference: (-0.0278678; 0.0718678)
Test for difference = 0 (vs not = 0): Z = 0.86 P-Value = 0.387

Fisher's exact test: P-Value = 0.476
```

Use this p-value when the requirements for the usual test are not fulfilled

Usual test: Uses the normal approximation to the binomial distribution; requires N>5 and (1-p)N>5

In any of the two cases, since the p-value is greater than 0.05 (usual critical value), it is not possible to conclude that the proportions are significantly different.

11

Comparison of More than Two Means: Analysis of Variance

11.1 ANOVA (Analysis of Variance)

Stat > ANOVA

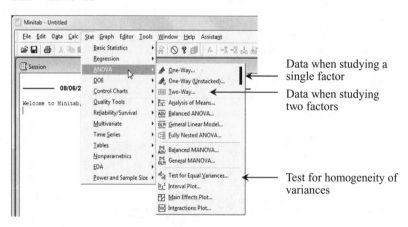

Data when studying a single factor

Data when studying two factors

Test for homogeneity of variances

11.2 ANOVA with a Single Factor

Example 11.1: The technicians of a paper processing company decide to carry out an experiment in order to study which variety of tree produces

Industrial Statistics with Minitab, First Edition. Pere Grima Cintas, Lluís Marco-Almagro and Xavier Tort-Martorell Llabrés.
© 2012 John Wiley & Sons, Ltd. Published 2012 by John Wiley & Sons, Ltd.

less contents of phenols in the waste of paper pulp. The following data represent the quantities obtained (in %).

Variety of tree A	1.9	1.8	2.1	1.8		
Variety of tree B	1.6	1.1	1.3	1.4	1.1	
Variety of tree C	1.3	1.6	1.8	1.1	1.5	1.1

Do any of the tree varieties produce more phenols?

This problem deals with the comparison of the means of three treatments, being 'Variety of tree' the only factor under study.

If we place the data for each type of tree in a different column, we have to proceed as shown below.

Stat > ANOVA > One-Way (Unstacked)

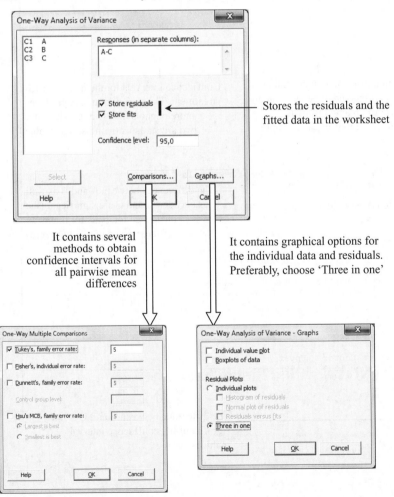

Stores the residuals and the fitted data in the worksheet

It contains several methods to obtain confidence intervals for all pairwise mean differences

It contains graphical options for the individual data and residuals. Preferably, choose 'Three in one'

If the data of the three varieties of trees came from the same normal distribution, a difference among their means like the one obtained or greater would only occur in 0.5% of all cases

```
One-way ANOVA: A; B; C

Source  DF      SS      MS     F      P
Factor   2  0.9000  0.4500  8.44  0.005
Error   12  0.6400  0.0533
Total   14  1.5400

S = 0.2309   R-Sq = 58.44%   R-Sq(adj) = 51.52%


                          Individual 95% CIs For Mean Based on
                          Pooled StDev
Level  N    Mean   StDev  ----+---------+---------+---------+-----
A      4  1.9000  0.1414                        (-------*--------)
B      5  1.3000  0.2121  (------*-------)
C      6  1.4000  0.2828     (------*------)
                          ----+---------+---------+---------+-----
                          1.20      1.50      1.80      2.10

Pooled StDev = 0.2309
```

Estimation of the population standard deviation (assumed to be the same in each group) obtained from the standard deviations of each sample

Confidence intervals for the mean of each treatment. The variety of tree A produces more phenols than B or C. The latter two are statistically indistinguishable

. . .

```
Tukey 95% Simultaneous Confidence Intervals
All Pairwise Comparisons
```

The differences between A and B as well as between A and C are significant (the confidence intervals for the mean differences do not include zero)

```
Individual confidence level = 97.94%


A subtracted from:

       Lower   Center   Upper   -----+---------+---------+---------+----
B    -1.0130  -0.6000  -0.1870  (---------*---------)
C    -0.8974  -0.5000  -0.1026     (---------*--------)
                                -----+---------+---------+---------+----
                                  -0.80     -0.40     -0.00      0.40


B subtracted from:

       Lower   Center   Upper   -----+---------+---------+---------+----
C    -0.2728   0.1000   0.4728                     (---------*--------)
                                -----+---------+---------+---------+----
                                  -0.80     -0.40     -0.00      0.40
```

The difference between B and C is not significant (its confidence interval includes zero)

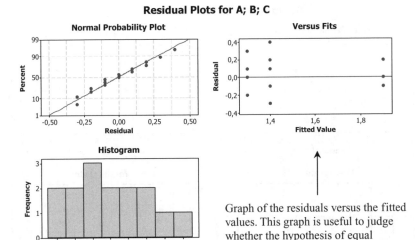

Residual Plots for A; B; C

Graph of the residuals versus the fitted values. This graph is useful to judge whether the hypothesis of equal variances holds.

Histogram and normal probability plot of residuals (useful to check for normality and to detect any anomalous values).

In the worksheet, both the residuals and the fitted values are stored (we checked these options in the initial dialog box).

↓	C1	C2	C3	C4	C5	C6	C7	C8	C9	C
	A	B	C	RESI1	RESI2	RESI3	FITS1	FITS2	FITS3	
1	1,9	1,6	1,3	-0,0	0,3	-0,1	1,9	1,3	1,4	
2	1,8	1,1	1,6	-0,1	-0,2	0,2	1,9	1,3	1,4	
3	2,1	1,3	1,8	0,2	0,0	0,4	1,9	1,3	1,4	
4	1,8	1,4	1,1	-0,1	0,1	-0,3	1,9	1,3	1,4	
5		1,1	1,5		-0,2	0,1		1,3	1,4	
6			1,1			-0,3			1,4	
7										

Worksheet 1 ***

If the data of the three varieties of trees are stored in one single column, we have to choose the following.

Stat > ANOVA > One-Way

All options and results are the same as before with one exception: Option 'Three in one' is substituted by 'Four in one' since a fourth graph is created representing the residuals as a function of the order they appear in the worksheet.

Residual Plots for Phenols

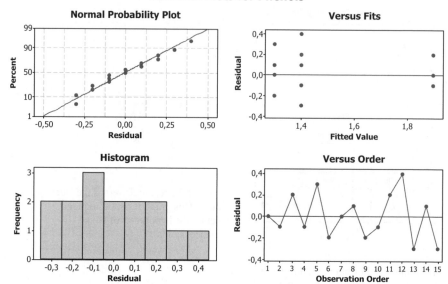

11.3   ANOVA with Two Factors

Example 11.2: A study is carried out to analyze whether the type of painting, among three types available, affects the hardness of a certain type of enamel. The firing of the parts is done in an oven with four trays, one above another, and with 15 parts on each. Since it can be assumed that the tray might have an influence on the result, because of its position in the oven, a blocked design is used putting five parts of each type of painting on each tray. The following hardness measures are obtained:

	Painting		
Tray	1	2	3
1	7.3 7.0 7.0 6.5 7.6	6.2 6.5 6.4 7.2 6.3	5.5 6.0 6.7 6.1 6.5
2	6.9 7.1 7.2 7.4 6.3	5.7 6.4 6.9 6.0 6.8	6.9 5.7 7.0 6.5 6.3
3	7.9 6.8 7.8 7.3 6.9	6.4 6.9 6.4 7.2 7.2	6.6 6.2 6.3 6.5 7.0
4	7.7 7.6 6.5 7.5 8.0	6.6 6.5 7.1 6.2 6.3	6.0 6.5 6.8 6.4 5.7

Can we affirm that the type of painting affects the degree of hardness?

Notice that now we have two factors which may affect the response: the type of painting and the tray on which the parts are placed.

Place obtained values in the worksheet as indicated below:

↓	C1	C2	C3
	Hardness	Painting	Tray
1	7,3	1	1
2	7,0	1	1
3	7,0	1	1
4	6,5	1	1
5	7,6	1	1
6	6,2	2	1
7	6,5	2	1
8	6,4	2	1
9	7,2	2	1
10	6,3	2	1
11	5,5	3	1
12	6,0	3	1
13	6,7	3	1
14	6,1	3	1
15	6,5	3	1
16	6,9	1	2
17	7,1	1	2
18	7,2	1	2
19	7,4	1	2
20	6,3	1	2
21	5,7	2	2
22	6,4	2	2
23	5,9	2	2

C1: Values of the hardness

C2: Values 1, 2 and 3 are used to identify the type of painting corresponding to each hardness value in column C1. These values can be introduced as follows: **Calc > Make Patterned Data > Simple Set of Numbers, From first value**: 1; **To last value**: 3; **In steps of:** 1; **Number of times to list each value**: 5; **Number of times to list the sequence**: 4

C3: Values from 1 to 4 are used to identify the tray. They can also be entered using the above mentioned utility to create patterned data.

Stat > ANOVA > Two-way

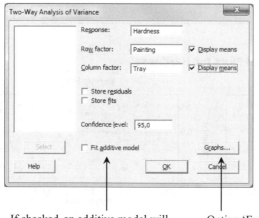

Display means: To show confidence intervals for the means

To store residuals and fitted values

If checked, an additive model will be fitted. That is, no interactions are considered

Option 'Four in one' summarizes all relevant information

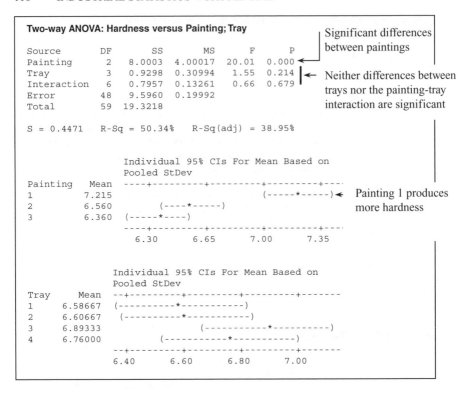

Two-way ANOVA: Hardness versus Painting; Tray

Source	DF	SS	MS	F	P
Painting	2	8.0003	4.00017	20.01	0.000
Tray	3	0.9298	0.30994	1.55	0.214
Interaction	6	0.7957	0.13261	0.66	0.679
Error	48	9.5960	0.19992		
Total	59	19.3218			

Significant differences between paintings ←

← Neither differences between trays nor the painting-tray interaction are significant

S = 0.4471   R-Sq = 50.34%   R-Sq(adj) = 38.95%

```
                     Individual 95% CIs For Mean Based on
                     Pooled StDev
Painting   Mean    ----+---------+---------+---------+---
1          7.215                              (-----*-----)←
2          6.560             (----*-----)
3          6.360    (-----*----)
                   ----+---------+---------+---------+---
                    6.30      6.65      7.00      7.35
```

Painting 1 produces more hardness

```
                     Individual 95% CIs For Mean Based on
                     Pooled StDev
Tray     Mean     --+---------+---------+---------+-------
1        6.58667  (----------*-----------)
2        6.60667  (----------*-----------)
3        6.89333              (-----------*----------)
4        6.76000        (-----------*-----------)
                  --+---------+---------+---------+-------
                  6.40      6.60      6.80      7.00
```

Residual Plots for Hardness

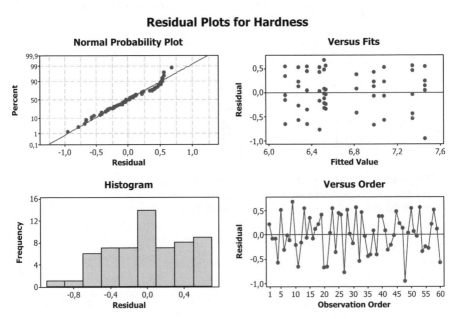

Example 11.3: The tensile strength of different pieces of clothes tinted by two different procedures (A and B) is measured. The experiment is carried out using pieces of cotton, nylon, and silk. The obtained data are:

	Treatment A	Treatment B
Cotton	25.5 24.1 24.8 23.7 23.8	29.0 29.8 29.4 28.0 29.7
Nylon	17.5 16.9 19.7 17.1 17.0	29.7 30.2 31.4 30.7 31.3
Silk	14.5 13.9 15.5 18.8 14.1	25.0 25.8 23.6 24.2 24.3

Are there significant differences between both treatments with respect to the tensile strength? And what about the materials?

Likewise in Example 11.2, proceed to enter the data in the worksheet and then go to the following.

Stat > ANOVA > Two-way

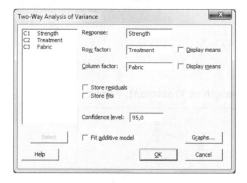

The following ANOVA table appears in the Session window:

Two-way ANOVA: Strength versus Treatment; Fabric

Source	DF	SS	MS	F	P
Treatment	1	611.175	611.175	474.12	0.000
Fabric	2	236.139	118.069	91.59	0.000
Interaction	2	85.275	42.638	33.08	0.000 ← The interaction is significant, too.
Error	24	30.937	1.289		
Total	29	963.526			

Unlike the previous example, there is an interaction between both factors. That means that the treatment effect depends on the fabric. That is, changing treatment A for treatment B does not have the same effect on all fabrics; it may represent an important advantage for one fabric, but may have no effect for another or may be even disadvantageous.

In presence of interaction, the changes observed in the response variable when changing one of the two factors depend on the value of the other.

Since there is an interaction, a graphical representation of the response's average values of both treatments is not of interest, because the mean difference observed will not be the same for each fabric.

Contrary to that, a useful graph to understand the data behaviour is the one obtained by clicking on button **Graph** and, thereafter, checking option **Individual value plot**.

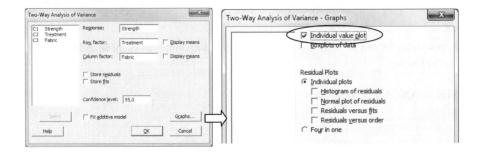

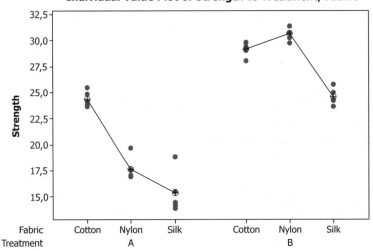

In this graph, we can observe, among other issues, how the change of treatment has a more important effect on the tensile strength of nylon than of cotton.

The lines do not show the same fabric profile for both treatments (they are not parallel) because the interaction fabric-treatment is significant.

11.4   Test for Homogeneity of Variances

Consider again the data on the comparison of phenol contents of different tree varieties. Can we affirm that there is a significant difference among the three varieties of trees with respect to the variability of phenol contents?

Stat > ANOVA > Test for Equal Variances.

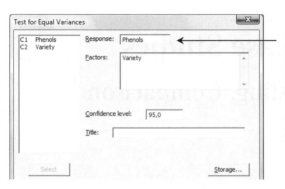

All data values of the response variable must be contained in the indicated column. In **Factors** enter the variable 'variety of tree', which indicates the tree variety corresponding to the phenol values in column C1.

The test results are summarized in the following graph:

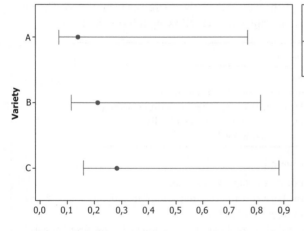

Bartlett: Assumes that the data come from normal distributions.

Levene: Assumes that the data come from any continuous distributions.

The confidence intervals greatly overlap and the p-values of both tests are large. Therefore, it cannot be stated that the variability is different among the three varieties of trees.

12

Part Two: Case Studies
Hypothesis Testing. Comparison of Treatments

12.1 Welding

To know more about the behaviour of an automatic welding procedure, 100 welds were made. The exact condition in which each weld was made was recorded. The welds were made in standard sheets and their shear stresses were tested. The file WELDING.MTW contains the data, with the following information:

Column	Content
C1	Test order
C2	Strength exerted by the electrodes
C3	Electric current measured with Bosch's ammeter
C4	Electric current measured with Miyachi's ammeter
C5	Shear breaking load

The questions to be answered are:

1. Is the exerted strength by the electrodes stable?
2. Is there a correlation between the exerted strength and the shear stress? Is there a cause and effect relation?
3. Can we say that the maximum shear stress is lower in the first 50 welds than in the last 50 welds?
4. The real value of the electric current varies from one weld to another (for various reasons it cannot remain constant). Can we say that the Bosch and Miyachi ammeters measure different?

Industrial Statistics with Minitab, First Edition. Pere Grima Cintas, Lluís Marco-Almagro and Xavier Tort-Martorell Llabrés.
© 2012 John Wiley & Sons, Ltd. Published 2012 by John Wiley & Sons, Ltd.

1. Start checking if the strength exerted by the electrodes has remained stable:

Graph > Time Series plot: Simple

Place the column C2 in **Series**, with all other options by default:

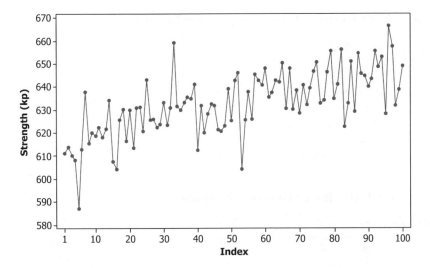

The strength has not remained stable. There is a tendency to increase.

2. Investigate if there is a correlation between strength and shear stress:

Graph > Scatterplot: Simple

Put the shear breaking load (C5) in Y, and strength (C2) X:

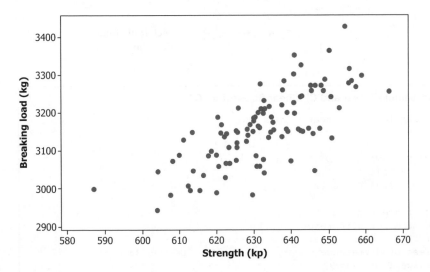

There is a correlation between breaking load and strength. The existence of a cause–effect relationship cannot be inferred from these data, but from technical considerations or experiments conducted.

3. Create an auxiliary column with 50 ones and 50 twos:

Calc > Make Patterned Data > Simple Set of Numbers

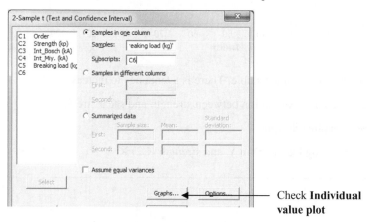

We now select: **Stat > Basic Statistics > 2-Sample t**

Check **Individual value plot**

Two-Sample T-Test and CI: Shear Stress (kg); C6

```
Two-sample T for Shear Stress (kg)

                      SE
C6   N    Mean  StDev  Mean
1    50  3114.1   81.4   12
2    50  3204.0   88.2   12

Difference = mu (1) - mu (2)
Estimate for difference: -89.9
95% CI for difference: (-123.6; -56.3)
T-Test of difference =0 (vs not =): T-Value =-5.30
P-Value=0.000 DF=97
```

The difference between the average of the top 50 values and the average of the last 50 is clearly significant (p-value = 0.000). Graphically:

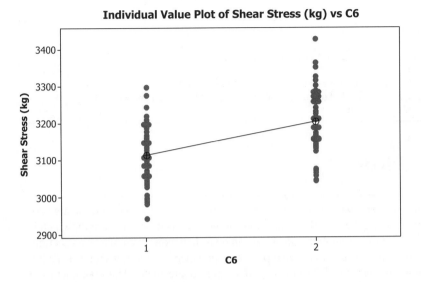

4.   Can we say that the Bosch and Miyachi ammeters measure different? Real current varies in each weld, so we must perform a paired data comparison of means.

Stat > Basic Statistics > Paired t

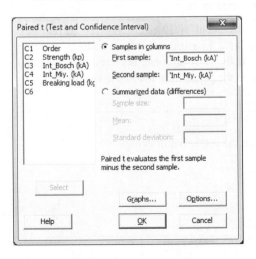

The result shows that the difference is clearly significant.

```
Paired T-Test and CI: Cur_Bosch (kA); Cur_Miy. (kA)

Paired T for Cur_Bosch (kA) - Cur_Miy. (kA)

                      N     Mean    StDev   SE Mean
Cur_Bosch (kA)      100  15.0050   0.2100   0.0210
Cur_Miy. (kA)       100  15.1030   0.2401   0.0240
Difference          100  -0.0980   0.1110   0.0111

95% CI for mean difference: (-0.1200; -0.0760)
T-Test of mean difference =0 (vs not =0): T-Value = -8.83
  P-Value =0.000
```

12.2 Rivets

An industry produces rivets for metal sheets used in the building sector. An essential feature, besides the dimensions of the rivet, is its shear strength, which is a measure of the strength that the rivet can resist in the direction of a cross section.

The intended use of a certain type of rivet and market conditions require having a minimum strength of 2500 psi (pounds per square inch). The manufacturing process gives this magnitude a variability that can be characterized by a normal distribution with $\sigma = 20$ psi.

The questions we pose are:

1. What should be the average value of strength if we accept 2.5% of the rivets with a value below the required minimum?

2. Some assembly requires the placement of 20 rivets. For the correct behaviour of the installation, at least 15 rivets must have a shear strength above the required minimum. If the rivets are manufactured with the process centred in 2525 psi, what is the probability that the assembly behaves correctly?

1. We have to find the value of μ in the following drawing:

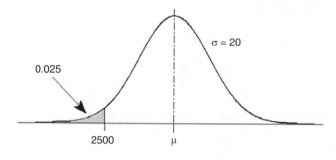

As the tail area is 2.5%, the value 2500 will be approximately 2σ far from μ, so μ will be around 2540. We can use Minitab to, by trial and error, find the exact value. Go to **Calc** > **Probability Distributions** > **Normal**:

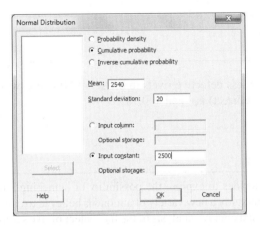

and the result is 0.0228. As it does not reach 2.5% we can try again reducing the average. But if we replace 2540 by 2539 in the above dialog box, we have a cumulative probability of 0.0256. As this last value is greater than 2.5%, the best option is keeping 2540 as the answer to this question.

2. In the same dialog box just showed, placing a mean of 2525 the probability of having values below 2500 is 0.1056.

   When placing 20 rivets and the probability that each of them is correct is constant, the number of correct rivets is a random variable that follows a binomial distribution. At least 15 correct rivets is the same as at most 5 defective rivets. Go to **Calc** > **Probability Distributions** > **Binomial**

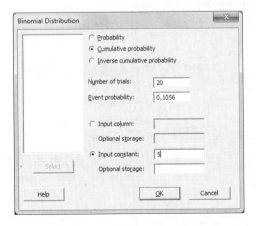

The result obtained is:

Cumulative Distribution Function

```
Binomial with n = 20 and p = 0.1056

x   P( X <= x )
5      0.985470
```

The probability of having 5 or less defective rivets in the set of 20 rivets (which is the same as having 15 or more correct) is equal to 0.9855.

12.3   Almonds

A nut producing company decides to explore the possibility of launching a new range of light almond. For this study a new variety of almond is believed to contain less fat than the commonly used one. An analysis of the fat content of five samples of both varieties is done to confirm the hypothesis. The results are:

Usual variety	27.0	26.9	27.3	27.2	27.1
New variety	26.8	27.0	26.9	27.1	26.8

Can we assure, with a significance level of 5%, that the new variety has less fat than the usual one?

This is a problem of comparison of means with independent samples. Go to **Stat** > **Basic Statistics** > **2-Sample t**

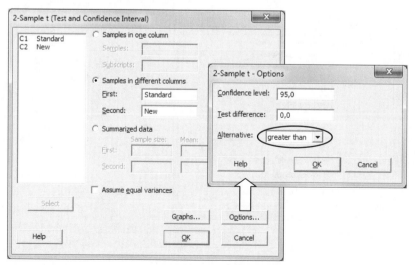

Choose 'greater than' as alternative hypothesis. That is, either the fat content is the same in both varieties, or – if the null hypothesis is rejected – the usual variety contains more fat. The result is:

```
Two-Sample T-Test and CI: Usual; New

Two-sample T for Usual vs New

        N    Mean   StDev   SE Mean
Usual   5  27.100   0.158    0.071
New     5  26.920   0.130    0.058

Difference = mu (Usual) - mu (New)
Estimate for difference: 0.1800
95% lower bound for difference: 0.0064
T-Test of difference = 0 (vs >): T-Value = 1.96
   P-Value = 0.045 DF = 7
```

The p-value obtained is 0.045. Therefore, with a significance level of 5% it can be said that the new variety has less fat than the usual variety.

12.4 Arrow

An arrow shaped plastic pieces are made with two injection moulding machines: A and B. Each machine has its own mould with 4 cavities, as shown in the following figure:

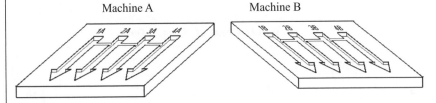

Data is available on ARROW.MTW file:

Column	Content
C1	Defects found in the parts produced by machine A
C2	As C1, but for machine B
C3	Tensile strength of parts made in 15 injections with machine A ($15 \times 4 = 60$ data)
C4	As C3 for machine B
C5	Cavity to which data from columns C3 and C4 belong.

The questions we pose are:

1. Looking at the distribution of defects in each machine, should the improvement measures be equally prioritized for both machines?

2. As for the tensile strength, is there a difference between machines? And between cavities?

1. Let's see what is the distribution of defects in each machine by:

Stat > Quality Tools > Pareto Chart

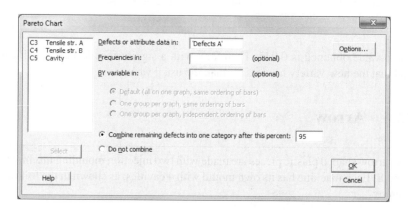

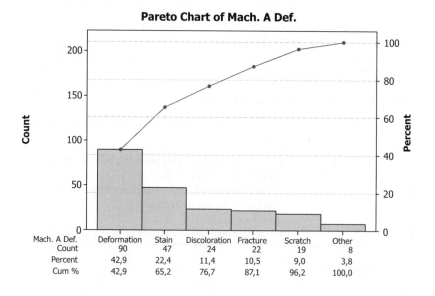

Pareto Chart of Mach. A Def.

Mach. A Def.	Deformation	Stain	Discoloration	Fracture	Scratch	Other
Count	90	47	24	22	19	8
Percent	42,9	22,4	11,4	10,5	9,0	3,8
Cum %	42,9	65,2	76,7	87,1	96,2	100,0

Doing the same for the machine B:

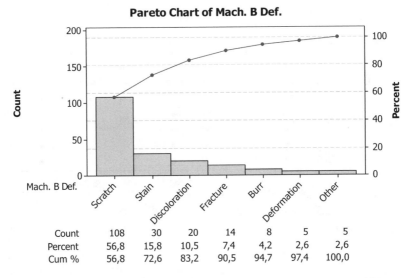

Pareto Chart of Mach. B Def.

Mach. B Def.	Scratch	Stain	Discoloration	Fracture	Burr	Deformation	Other
Count	108	30	20	14	8	5	5
Percent	56,8	15,8	10,5	7,4	4,2	2,6	2,6
Cum %	56,8	72,6	83,2	90,5	94,7	97,4	100,0

Assuming that the data are representative of both machines, it would be appropriate to focus on the problem of deformation in machine A and scratches in machine B.

2. First of all, a graphical analysis to compare machines and cavities can be performed:

Graph > Individual Value Plot: Multiple Y's With Groups

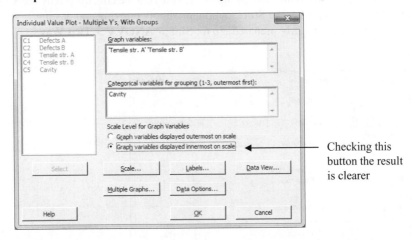

Checking this button the result is clearer

The graph shows no difference between machines, but more tensile strength than the rest is produced in cavity 3.

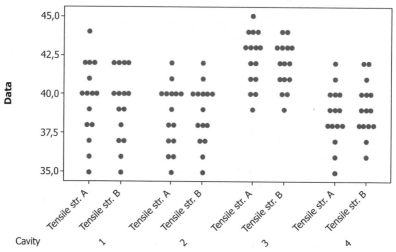

To apply the technique of analysis of variance it is necessary to have the responses in a single column, with another column indicating which machine has produced each value, and a third column indicating which cavity corresponds to each value.

The strength values can be placed in a single column with **Data > Stack > Columns,** using **Store subscripts in** to create the column indicating which machine corresponds to each value. The column of cavities can be constructed by stacking the column C5 twice, using the **Stack** option, or simply copying and pasting.

Once data is properly prepared, go to: **Stat > ANOVA > Two-Way**

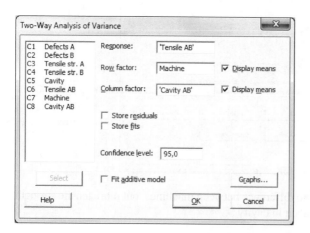

```
Two-way ANOVA: Resistance AB versus Machine; Cavity AB

Source        DF      SS       MS       F      P
Machine        1    0.033   0.0333   0.01   0.926
Cavity AB      3  220.867  73.6222  18.96   0.000
Interaction    3    3.767   1.2556   0.32   0.808
Error        112  434.800   3.8821
Total        119  659.467

S = 1.970   R-Sq = 34.07%   R-Sq(adj) = 29.95%

                          Individual 95% CIs For Mean Based on
                          Pooled StDev
Machine           Mean   --+---------+---------+---------+-------
Mach. A Res.   39.7833   (----------------*----------------)
Mach. B Res.   39.7500   (----------------*----------------)
                         --+---------+---------+---------+-------
                         39.30     39.60     39.90     40.20

                          Individual 95% CIs For Mean Based on
Cavity                    Pooled StDev
AB                Mean   -------+---------+---------+---------+--
1              39.4333          (----*----)
2              38.6667   (----*----)
3              42.0667                     (---*----)
4              38.9000      (---*----)
                         -------+---------+---------+---------+--
                         39.0      40.5      42.0      43.5
```

Annotations (right side of ANOVA): There is no difference between machines. There is a difference between cavities. No cavity-machine interaction (the effect of the machine does not depend on the cavity and viceversa)

Cavity 3 gives greater resistance in the two machines

Using the **Graph** button graphics of residuals versus fitted values can be made, but they do not show anything relevant in this occasion.

12.5   U Piece

A mechanism comprising a U-shaped part is manufactured. Its critical dimension is the width of the mouth, X. These parts, before being assembled in the mechanism, are subjected to a heat treatment in a furnace to increase its hardness. The oven is similar to a domestic oven (perhaps slightly higher) and the parts are placed in 5 trays, one above the other, supported in a two side support.

The nominal value of X is 12 ± 0.1 mm. We want to know if this heat treatment produces a systematic change of this value. To conduct the study the value of X before and after the treatment is measured for 100 parts. The number of the tray where each part has been placed is also registered.

The data are in the file U_PIECE.MTW. What conclusions can be drawn?

To get a first idea we create a new column with the difference: After-Before and we draw a dotplot:

Graph > Dotplot > Simple

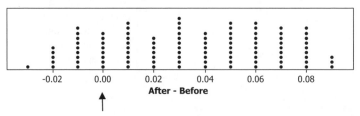

Notice that 0 appears off-centred and in most cases the heat treatment increases the value of this dimension. To confirm the influence we perform a comparison of means test for paired data:

Stat > Basic Statistics > Paired t,
First sample: Before, **Second Sample**: After

```
Paired T-Test and CI: Before; After
Paired T for Before - After

                N      Mean     StDev   SE Mean
Before        100   11.9969    0.0252    0.0025
After         100   12.0308    0.0388    0.0039
Difference    100   -0.03390  0.03216   0.00322

95% CI for mean difference: (-0.04028; -0.02752)
T-Test of mean difference=0 (vs not = 0): T-Value=-10.54
   P-Value=0.000
```

The heat treatment increases the X dimension 0.03 mm on average and this difference is clearly significant. At a first sight, it seems reasonable producing the pieces with a nominal value of 11.97 mm to reach the nominal value after the treatment. Let's see if the tray has some influence on the change of X. Go to:

   Stat > ANOVA > One-way

Response: After – Before (column we have created); **Factor:** Tray

```
One-way ANOVA: After-Before versus Tray

Source  DF        SS        MS       F      P
Tray     4   0.024594  0.006148   7.51  0.000
Error   95   0.077785  0.000819
Total   99   0.102379

S = 0.02861    R-Sq = 24.02%    R-Sq(adj) = 20.82%


                             Individual 95% CIs For Mean Based on
                             Pooled StDev
Level   N     Mean    StDev  ------+---------+---------+---------+---
1      20  0.02000  0.02384  (-----*-----)
2      20  0.02150  0.02996  (------*-----)
3      20  0.03300  0.03450      (------*-----)
4      20  0.03150  0.03133    (------*-----)
5      20  0.06350  0.02134                        (------*-----)
                             ------+---------+---------+---------+---
                              0.020     0.040     0.060     0.080
Pooled StDev = 0.02861
```

The increase of X in the fifth tray is greater than in the others. One might think of not using the fifth tray to reduce variability in the increase of X.

12.6   Pores

In a process of steel plates welding some pores appear in the weld. These pores are mainly due to the protective coating (to prevent oxidation) of the bearing plates. When burning the paint some gases are generated and are trapped in the weld; this creates some small bubbles that form pores.

Three different types of paint can be used. Each type of paint provides the same protection, but they are probably different in creating pores. Therefore, a test is performed. The test consists in welding five pairs of plates, each with one type of paint. As unpainted metal also produces pores when welded (due to soiling or other factors), only half of the plates are painted. Later, the porosity in the weld corresponding to the painted and the unpainted area are measured.

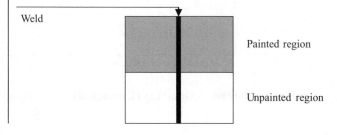

Paint type	A		B		C	
Region	Painted	Unpainted	Painted	Unpainted	Painted	Unpainted
Results	15	8	10	7	16	8
	21	6	13	9	22	8
	18	6	12	4	21	3
	17	7	11	6	17	5
	16	5	11	6	23	4

Is one type of paint better than the others?

The interesting outcome is the difference between the porosity in the painted and the unpainted areas, as this is the value of porosity due to the painting. These differences are calculated in the datasheet:

	C1	C2	C3	C4	C5	C6	C7	C8	C9	C10	C1
	Paint A	No/Paint A	Paint B	No/Paint B	Paint C	No/Paint C		A (c1-c2)	B (c3-c4)	C (c5-c6)	
1	15	8	10	7	16	8		7	3	8	
2	21	6	13	9	22	8		15	4	14	
3	18	6	12	4	21	3		12	8	18	
4	17	7	11	6	17	5		10	5	12	
5	16	5	11	6	23	4		11	5	19	
6											

Worksheet 1 ***

A first exploratory data analysis (**Graph > Individual Value Plot > Multiple Y's > Simple**) shows that the average value of porosity produced by B tends to be lower than the other two paintings.

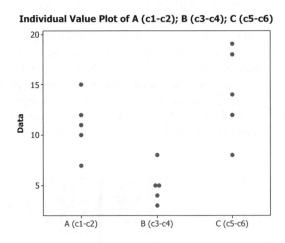

Individual Value Plot of A (c1-c2); B (c3-c4); C (c5-c6)

By analysis of variance: **Stat > ANOVA > One-Way (Unstacked)**

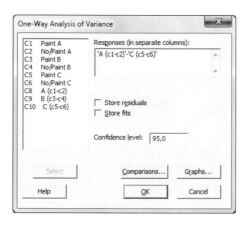

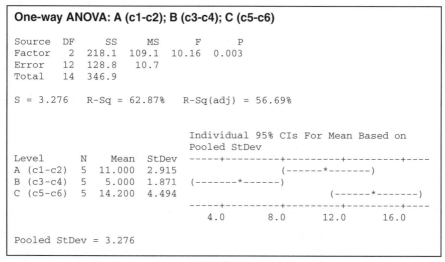

The difference is statistically significant (p-value = 0.003). Paint B produces less porosity.

When scatterplots of porosity in the painted area versus the unpainted area are drawn, no relationship is observed. Therefore, one could perform the analysis of variance with data of porosity in the painted part without subtracting the porosity in the unpainted part. Done in this way, the difference in favour of B is even clearer (F = 17.23, p-value = 0.000).

Part Three

MEASUREMENT SYSTEMS STUDIES AND CAPABILITY STUDIES

When measuring a characteristic of a product it is sometimes forgotten that the whole measurement process is subject to variability. There are no perfect measuring devices, nor do people involved in the measuring process (operators) always behave in the same way. The result is that, inevitably, the available data are contaminated by the variability introduced by the measurement system.

The good news is that we can design a data collection plan that actually allows separating the variability due to parts of that introduced by the operators and the measuring device. These studies are called: repeatability studies (variability of the device) and reproducibility studies (variability introduced by metering or by environmental factors). Minitab is helpful for this type of study and Chapter 13, the first chapter of Part 3, is dedicated to that issue.

The next two chapters (14 and 15) are devoted to capability studies. They are aimed at measuring the natural variability of a process when it is solely affected by what are called 'random variability causes'. They are characterized by being many, inevitable, small and almost imperceptible one by one. However, when acting together, they provoke this inevitable variability that we call capacity.

Two types of capability are of interest: short and long term. Short-term capability is measured in a short space of time, generally by measuring units produced in a row, one after another; long-term capability is measured over a long period of production and this gives the opportunity to variability causes such as changes in shift, raw materials, environmental conditions, state of the machine, etc. to appear. Sometimes short-term capability is called 'machine capability' because it may refer only to the variation produced in a particular machine, while long-term is referred as process capability because it includes variation from a whole process formed by the union of several machines.

As has just been said, each capability type, in principle, requires a sample taken in a different way; however, both capabilities can be measured using the same sample.

This is accomplished by sampling 4 or 5 consecutive units (number can vary) with some frequency over a long period of time; long enough to give the opportunity for the above-mentioned causes to appear. When data is collected in this way, the short-term capability is given by the variability within samples – since they are consecutive units, it is assumed that they are all produced under the same conditions and thus only subjected to random variability causes. The long-term capability is measured from the overall variability of all data.

Obviously the variability of a process is acceptable or not, depending on the tolerances of the product being manufactured. The acceptable variability is very different in a process filling sacks of cement of 50 kg than in a process filling instant coffee cans of 200 g, as a negligible variation in the first case can be quite intolerable in the second. Capability indexes measure the relationship between product tolerances and process capability. The best known relates the tolerance interval with the width of the process variation. Minitab calls Cp, according to the usual notation, the short-term capability index and Pp the long-term one.

The width of the process is usually defined as six times the process standard deviation (this is the default used by Minitab). This means that when the index (Cp or Pp) is 1 there are 0.3% of the units produced outside the tolerance interval. Since this is a significant number, which in many cases cannot be ignored, a solution is to take eight sigmas for the width of the variability ($\pm 4\sigma$ instead of $\pm 3\sigma$). Another solution is to require an index of 1 when working with 8σ, and of 1.33 when using 6σ (both requirements are equivalent).

The Cp (or Pp) compares tolerances and process variability, but does not take into account whether or not the process is on target. To account for this aspect a new index Cpk (or Ppk) is defined. It is calculated to be equal to Cp if the process is perfectly on target, and gets smaller as the process mean moves away from target. When the product only has one tolerance limit it is possible to calculate Cpk but not Cp.

In addition to the above-mentioned indexes Minitab provides a statistical summary of data including the standard deviation values in the short (within) and long (overall) term, the proportion of defects observed in the sample and the expected proportion calculated from the normal distribution with parameters estimated from the data. An interesting feature is provided by the Capability Sixpack™ option which, in addition to the capability study key information, shows control charts to see if the process is stable ('in state of control') and a normality test. It also offers the possibility of conducting capability analysis when the data is not normal. In this case, several distributions can be used: Weibull, lognormal, extreme values, etc.

13

Measurement System Study

13.1 Crossed Designs and Nested Designs

Crossed designs are those in which each operator measure all parts several times. These are the most common designs, especially if dimensional features are measured.

Scheme of the crossed design (all operators measure all parts)

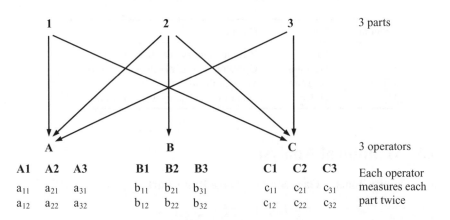

A1	**A2**	**A3**	**B1**	**B2**	**B3**	**C1**	**C2**	**C3**	
a_{11}	a_{21}	a_{31}	b_{11}	b_{21}	b_{31}	c_{11}	c_{21}	c_{31}	
a_{12}	a_{22}	a_{32}	b_{12}	b_{22}	b_{32}	c_{12}	c_{22}	c_{32}	

In a nested design, however, each part is measured by only one operator; being the type of design used in destructive tests. In nested designs, each operator should also measure each part several times, but in this case this is clearly not possible. Thus, an alternative solution is that each operator takes measures of several parts, the more alike as possible (i.e. parts produced under the same conditions); such that the heterogeneity of these parts is negligible and, hence, can practically be considered as a single part.

Industrial Statistics with Minitab, First Edition. Pere Grima Cintas, Lluís Marco-Almagro and Xavier Tort-Martorell Llabrés.
© 2012 John Wiley & Sons, Ltd. Published 2012 by John Wiley & Sons, Ltd.

Nested design (each operator measures only some of the parts):

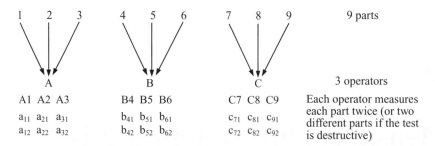

1 2 3	4 5 6	7 8 9	9 parts
A	B	C	3 operators
A1 A2 A3	B4 B5 B6	C7 C8 C9	Each operator measures
a_{11} a_{21} a_{31}	b_{41} b_{51} b_{61}	c_{71} c_{81} c_{91}	each part twice (or two
a_{12} a_{22} a_{32}	b_{42} b_{52} b_{62}	c_{72} c_{82} c_{92}	different parts if the test is destructive)

13.2   File 'RR_CROSSED'

The file RR_CROSSED.MTW contains the data needed to carry out an R&R study in which three operators take measures of ten different parts, three measures of each part. Each operator measures the ten parts consecutively in a random order, without previous knowledge of the part that is being measured. This measuring procedure is followed by each one of the three operators. The data file content is organized as follows:

Column	Name	Content
C1	Part	Part identifier (values from 1 to 10)
C2	Operator	Operator identifier (values from 1 to 3)
C3	Measure	Value of the measure
C4	Order	Order in which parts were measured

13.3   Graphical Analysis

Stat > Quality Tools > Gage Study > Gage Run Chart

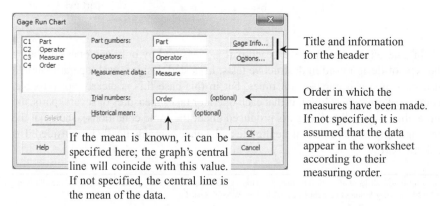

Title and information for the header

Order in which the measures have been made. If not specified, it is assumed that the data appear in the worksheet according to their measuring order.

If the mean is known, it can be specified here; the graph's central line will coincide with this value. If not specified, the central line is the mean of the data.

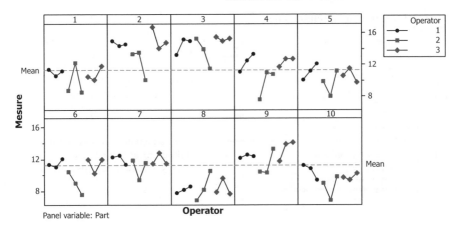

Panel variable: Part   **Operator**

Conclusions that can be drawn from this graph:

- There are differences between the parts. For example, parts 2 and 3 show values larger than the mean, whereas the values of parts 8 and 10 lie below the mean.

- Worker 2 measures with higher variability than the others. In addition, the values of operator 2 are generally smaller than those of operators 1 and 3.

Notice that the graph obtained is the same whether the order of the measures is specified or not, because the data are already sorted according to the measuring order.

13.4   R&R Study for the Data in File 'RR_CROSSED'

In this case, the adequate approach is a crossed design since each operator measures all parts several times. Thus, go to: **Stat > Quality Tools > Gage Study > Gage R&R Study (Crossed)**

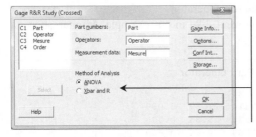

ANOVA: Based on the analysis of variance. This is the exact method which requires tedious calculations if done by hand.

Xbar and R ($\bar{X} - R$ graph): Approximate method, which is used to avoid the calculations of the exact method.

With Minitab, the use of ANOVA is recommended, which is actually the default option. The Xbar and R method is only advisable for comparison with other studies conducted with this method.

Options, with all default values:

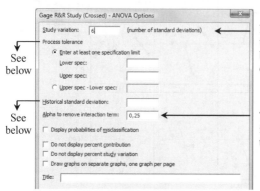

See below → (arrow pointing to Process tolerance section)

See below → (arrow pointing to Alpha to remove interaction term section)

Number of standard deviations that define the amplitude of the bell-shaped curves that represent each of the sources of variation

The interaction term is excluded if the corresponding p-value in the ANOVA table exceeds 0.25

Process tolerance refers to the tolerance interval for the measured magnitude. If this value is specified, Minitab displays which percentage of the amplitude is due to the variation of the measurement system and which percentage is due to each of its components. Furthermore, if the corresponding option is activated, the probabilities of erroneous classification are shown (seen later on).

Historical standard deviation. If the historical standard deviation of the measured magnitude is known, it can be specified here. Minitab will then show the percentage of that total variability due to the measurement process and each of its components.

The results shown in the Session window can be divided into three parts: (1) tables of the analysis of variance; (2) table of the variance components; (3) table of the study of variation.

13.4.1   Tables of the Analysis of Variance

These analyze whether differences exist between parts, between operators or whether a part-operator interaction exists. The displayed table is the same as the obtained by going to: **Stat > ANOVA > Two-Way.** There, in **Response** enter Measure; in **Row factor**: Part, and finally in **Column factor**: Operator.

If the interaction term is not significant, like in our example, Minitab shows a second ANOVA table which excludes it.

Two-way ANOVA: Measure versus Part; Operator

Source	DF	SS	MS	F	P
Part	9	286.033	31.7814	21.37	0.000
Operator	2	45.635	22.8173	15.35	0.000
Interaction	18	17.261	0.9589	0.64	0.849
Error	60	89.217	1.4869		
Total	89	438.145			

13.4.2   Table of Variance Components

This following table is not essential, because the information contained therein can be inferred from the third table (on the study of the variation).

Gage R&R		
		%Contribution
Source	VarComp	(of VarComp)
Total Gage R&R	2.08017	38.10
Repeatability	1.36510	25.00
Reproducibility	0.71507	13.10
Operator	0.71507	13.10
Part-To-Part	3.37959	61.90
Total Variation	5.45976	100.00

Observe that $2.08017 = 1.36510 + 0.71507$ and $5.45976 = 2.08017 + 3.37959$. Likewise, the percentage contribution can be inferred as follows: $38.10 = (2.08017/5.45976) \times 100$, etc.

13.4.3   Table of the Study of Variation

The column named 'StdDev (SD)' contains the square roots of column 'VarComp' presented in the previous table. That is, $1.44228 = \sqrt{2.08017}$, etc.

'Study Var' contains the values of the previous column multiplied by 6. These values represent the amplitude of the bell-shaped curves that correspond to each of the sources of variation, and which contain 99.7% of all observations (6σ is the default value).

'%Study Var' represents the percentage corresponding to the amplitude of each source of variation with respect to total variation. For example: $61.73 = (8.6537/14.0197) \times 100$.

		Study Var	%Study Var
Source	StdDev (SD)	(6 * SD)	(%SV)
Total Gage R&R	1.44228	8.6537	61.73
Repeatability	1.16838	7.0103	50.00
Reproducibility	0.84562	5.0737	36.19
Operator	0.84562	5.0737	36.19
Part-To-Part	1.83837	11.0302	78.68
Total Variation	2.33661	14.0197	100.00

13.4.4   Scheme of the Sources of Variation

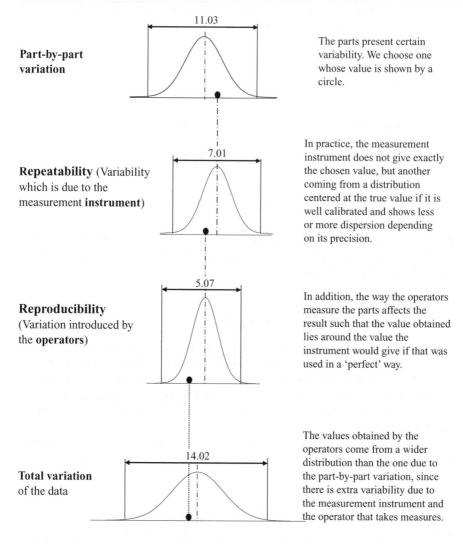

Part-by-part variation

11.03

The parts present certain variability. We choose one whose value is shown by a circle.

Repeatability (Variability which is due to the measurement **instrument**)

7.01

In practice, the measurement instrument does not give exactly the chosen value, but another coming from a distribution centered at the true value if it is well calibrated and shows less or more dispersion depending on its precision.

Reproducibility (Variation introduced by the **operators**)

5.07

In addition, the way the operators measure the parts affects the result such that the value obtained lies around the value the instrument would give if that was used in a 'perfect' way.

Total variation of the data

14.02

The values obtained by the operators come from a wider distribution than the one due to the part-by-part variation, since there is extra variability due to the measurement instrument and the operator that takes measures.

Notice that the amplitude of the last bell-shaped curve, representing the total variation, is not the sum of the amplitudes of the other three bell-shaped curves. For this reason, the values in column %Study Var that correspond to Total Gage R&R and Part-To-Part do not sum up 100.

13.4.5   Graphical Window

Using all default options, the graphical window has the following appearance:

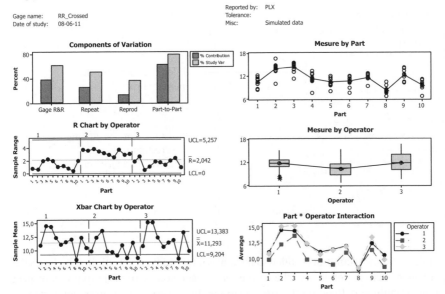

Gage R&R (ANOVA) for Mesure

	Components of	Graphical representation of each component's contribution
Variation:		according to the analysis of the variance components and the study of the variation.

Components of
Variation:

Graphical representation of each component's contribution according to the analysis of the variance components and the study of the variation.

R and Xbar Chart
by Operator

In the graph of the ranges (**R Chart by Operator**), it is observed that a larger variability is present in the measures made by the second operator than in the ones taken by the other two operators. The points beyond the control limits (**Graph: Xbar Chart by Operator**) may be interpreted in the sense that the measurement system is able to identify the differences between the different parts.

Measure by Part

Clearly, a part-by-part variation is observed.

Measure by
Operator

Operator 2 presents the lowest mean value.

Part * Operator
Interaction

The differences between parts can be clearly observed, as well that the second operator presents lower values than those of the other two operators. No differences are observed between the first and the third operator. If interaction does not exist, these lines are approximately parallel.

13.4.6 Comparison with the Part's Tolerance

In **Options,** one can specify the amplitude of the tolerance interval.

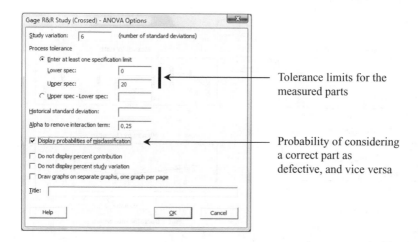

Tolerance limits for the measured parts

Probability of considering a correct part as defective, and vice versa

In this case, a new column is added to the table of the study of variation. It includes the percentages corresponding to the amplitude of each source of variation with respect to the amplitude defined by the specifications. In general, a measurement system is considered correct if the amplitude defined by its variability (Total Gage R&R) is less than a 20% of the specification's interval. As shown below, it is not the case in our example.

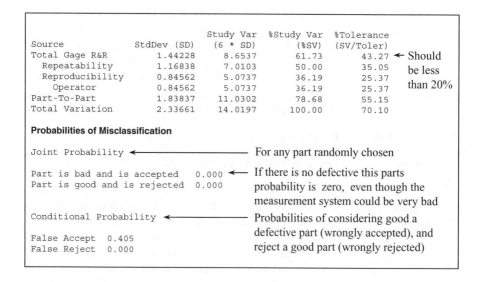

		Study Var	%Study Var	%Tolerance
Source	StdDev (SD)	(6 * SD)	(%SV)	(SV/Toler)
Total Gage R&R	1.44228	8.6537	61.73	43.27 ← Should
Repeatability	1.16838	7.0103	50.00	35.05 be less
Reproducibility	0.84562	5.0737	36.19	25.37 than 20%
Operator	0.84562	5.0737	36.19	25.37
Part-To-Part	1.83837	11.0302	78.68	55.15
Total Variation	2.33661	14.0197	100.00	70.10

Probabilities of Misclassification

Joint Probability ◄─────────────── For any part randomly chosen

Part is bad and is accepted   0.000 ◄── If there is no defective this parts
Part is good and is rejected 0.000   probability is zero, even though the
                                           measurement system could be very bad

Conditional Probability ◄─────────── Probabilities of considering good a
                                           defective part (wrongly accepted), and
False Accept   0.405   reject a good part (wrongly rejected)
False Reject   0.000

These probabilities may be wrongly interpreted. In general, the aforementioned criterion is used: a measurement system is considered correct if its variability is less than 20% (in some cases less than 10% is used) of the product's tolerance interval.

When product specifications are introduced, a new bar showing the relation (in percentage) between the process variation width and the tolerance interval width appears in the **Components of Variation** graph.

13.5   File 'RR_NESTED'

The file RR_NESTED.MTW contains the data needed to carry out an R&R study in which 12 parts and 3 operators are used. The parts are divided into 3 groups, each of 4 units. Each operator randomly measures three times all parts in one group, without previous knowledge of the part that is being measured. The data file content is organized as follows:

Column	Name	Content
C1	Part	Part identifier (values from 1 to 12)
C2	Operator	Operator identifier (A, B, or C)
C3	Measure	Value of the measure

13.6   Gage R&R Study for the Data in File 'RR_NESTED'

In this case, the adequate approach is a nested design since each operator measures only a subset of the parts. Thus, go to: **Stat > Quality Tools > Gage Study > Gage R&R Study (Nested)**.

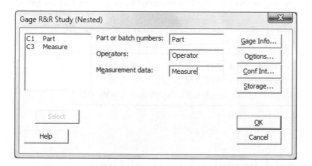

Gage Info and **Options:** Clicking on them open up identical dialog boxes to those produced in case of a crossed design. Likewise, the output displayed in the session window has the same structure as that of previous designs.

With these data, the following graph is constructed:

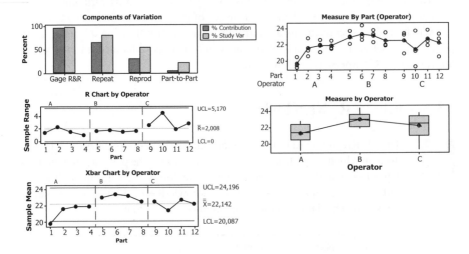

It can be observed that the variability between parts (Part-by-Part) is very small as compared with the one of the measurement system (Gage R&R), which indicates a nonappropriate measurement system.

13.7   File 'GAGELIN'

This file, included in the Minitab sample data folder, is used to illustrate how to carry out a linearity study in the context of measurement systems. Five parts of different sizes are taken representing the expected range of the measurements. Each part is measured by a pattern system to determine its reference (master) value. Then, an operator takes twelve measures of each part, using the system under study. The file content is organized as follows:

Column	Name	Content
C1	Part	Number of the part
C2	Master	Master value of the part
C3	Response	Measurement result with the system under study

13.8   Calibration and Linearity Study of the Measurement System

Retrieve the worksheet file: GAGELIN.MTW and do the following.

Stat > Quality Tools > Gage Study > Gage Linearity and Bias Study

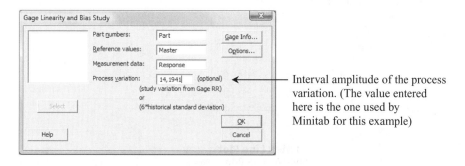

Interval amplitude of the process variation. (The value entered here is the one used by Minitab for this example)

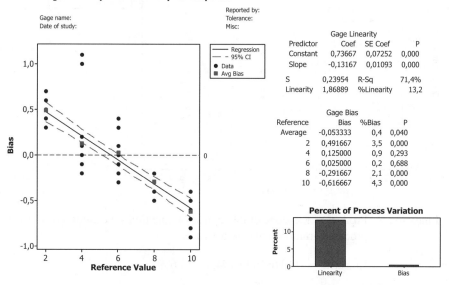

Where Does the Output Information of the Linearity Study Come from and How to Interpret It?

Graph: Shows the difference between the measured value and the real one (**Bias**), versus the real value (**Reference**). It may also be drawn by placing the difference between C3 (Response) and C2 (Master) in column C4 (named Difference) and then choosing:

Stat > Regression > Fitted Line Plot

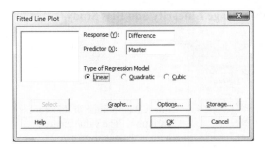

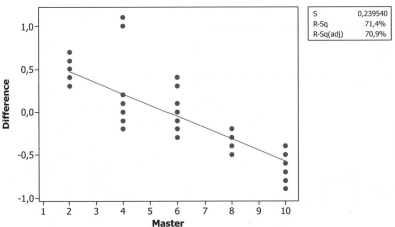

Fitted Line Plot
Difference = 0,7367 - 0,1317 Master

Linearity:	Product of the slope of the regression line times the amplitude of the process variation. Thus, in our case:

Linearity = 0.13167×14.1941 = 1.8689.

%Linearity	Slope of the regression line multiplied by 100. Here, the variation due to linearity is equivalent to 13% of the total variation.

Bias:	In the first line, the average of all differences between the measured and reference values is shown (corresponding in this case to the mean of column C4). Just beneath, the corresponding deviations to each reference value are shown.

%Bias:	%Bias = (|Bias|/process width)×100 = (-0.053/14.1941)×100 = 0.3757. This means that the bias introduced by the measurement system is approximately a 0.4% of the total variation.

14

Capability Studies

14.1 Capability Analysis: Available Options

Take a look of the available options to carry out a capability study.

Stat > Quality Tools. . .

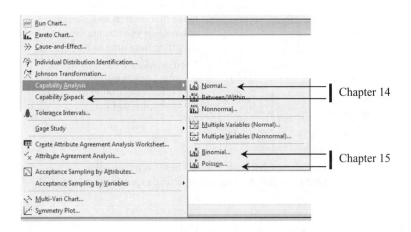

The present chapter focuses on capability studies for process variables following a Normal distribution. To carry out such capability studies you can use two options: **Capability Analysis** or **Capability Sixpack**. The next chapter deals with capability studies for binomial and Poisson distributions.

Industrial Statistics with Minitab, First Edition. Pere Grima Cintas, Lluís Marco-Almagro and Xavier Tort-Martorell Llabrés.
© 2012 John Wiley & Sons, Ltd. Published 2012 by John Wiley & Sons, Ltd.

14.2   File 'VITA_C'

Pharmaceutical companies must fulfil the European Pharmacopoeia standards that among other requirements dictate that the probability that the weight of a pill deviates at least 5% from its target value must be less than 0.5%.

To carry out a capability study of the production process of pills with target value of 3.25 g., samples of 5 units are taken every 15 minutes during 10 hours. The file VITA_C.MTW contains measures corresponding to the weights of 200 sampled pills. Our purpose is to find out if the process is capable to meet this requirement.

14.3   Capability Analysis (Normal Distribution)

Stat > Quality Tools > Capability Analysis > Normal

The data in column 'Weight'

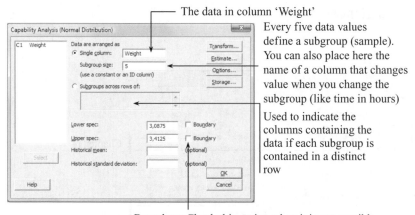

Every five data values define a subgroup (sample). You can also place here the name of a column that changes value when you change the subgroup (like time in hours)

Used to indicate the columns containing the data if each subgroup is contained in a distinct row

Boundary: Check this option when it is not possible (due to a control) to have units outside specification limits

14.4   Interpreting the Obtained Information

The graphical window displayed below is obtained. In the following, you will find a detailed description of the results obtained:

① Process data: Lower and upper specification limits (**LSL** and **USL**, you must enter at least one). **Target**: shown if introduced previously (optional). Global data mean (**Sample Mean**), total number of data values (**Sample N**), standard deviations estimated via the variability within the samples (**StDev (Within)**) and via the global variability (**StDev(Overall)**).

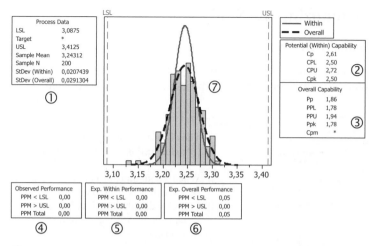

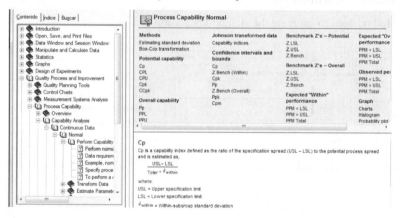

② Short time capability indexes (also called 'machine capability'): These indexes are computed from the variation within the samples. To see the detailed formulas, click on **Help** in the dialog window **Capability Analysis (Normal Distribution)**, then click on the option: **see also** (top right of the displayed window) and finally choose the topic: **Methods and formulas**.

③ Similar to item two, but based on the global variability (long-term capability or process capability).

④ Number of values in parts per million (PPM) that are observed outside (below and above) the specification limits.

⑤ Expected PPM outside specifications obtained based on the variation within samples. This theoretical value that reflects the variability is obtained from the normal distribution, centred at the sample mean.

⑥ Similar to the above, but based on the global or total variation.

⑦ Histogram of the data overlaid with two Bell-shaped curves representing, respectively, the theoretic global variation (overall, dash line) and an assumed minimum achievable variability (within, solid line), if the process remains stable over time.

14.5   Customizing the Study

14.5.1   Transform

When the original data are not normal, you may use a simple mathematical transformation to obtain data that, for practical purposes, are normally distributed.

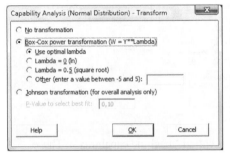

The default option is: No **transformation**

When a data transformation is required, it is typical to use the Box-Cox distribution. Lambda = 0 is the widely used logarithmic transformation.

14.5.2   Estimate

Method used to estimate the standard deviation.

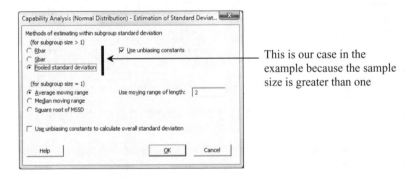

This is our case in the example because the sample size is greater than one

Usually, there is no need to change the default options. The within standard deviation is estimated via the average of the sample variances (if the sample sizes differ, each variance is weighted by the degrees of freedom of each sample).

It is well known that the sample variance s^2 is an unbiased estimator of the population variance σ^2; however, the sample standard deviation s is not an unbiased estimator of the population standard deviation σ. To make it unbiased, it is necessary to apply a bias correction factor which is a constant that depends on the degrees of freedom used to estimate s^2. Minitab applies it by default (**use unbiasing constants**).

When the subgroup size is one, by default the short time variability is computed via the average moving range of order two.

14.5.3 Options

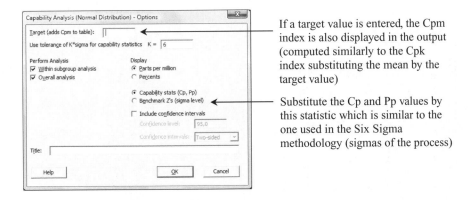

If a target value is entered, the Cpm index is also displayed in the output (computed similarly to the Cpk index substituting the mean by the target value)

Substitute the Cp and Pp values by this statistic which is similar to the one used in the Six Sigma methodology (sigmas of the process)

Sometimes, 8σ is used as the width of the bell-shaped curve to compute the capability indexes.

To require Cp = 1 using 8σ as the width of the bell-shaped curve is the same as requiring Cp = 1.33 using 6σ.

14.5.4 Storage

Stores the chosen values in the first empty available columns of the worksheet.

14.6 'Within' Variability and 'Overall' Variability

14.6.1 What Are They?

Within refers to the within samples variation. It gives an idea of the variability observed within a short period of time (also called 'short-term' variability) without taking into account causes like raw material changes, machine fouling, operator shifts, etc. that will inevitably appear. It can be interpreted as the minimal variability that can be achieved if all causes affecting the process along time were eliminated.

Overall refers to the global variability of the data, also called 'long-term' variability. It is the real variability of the production process.

14.6.2   How Are They Computed?

↓	C1	C2
	Sample	Mean
1	1	2
2	1	4
3	1	5
4	1	6
5	2	12
6	2	13
7	2	14
8	2	15
9	3	6
10	3	7
11	3	8
12	3	10
13		

Worksheet 1 ***

Let's use the simple set of numbers displayed on the right to illustrate the difference between both types of variability and to show how to compute them.

We have only three subgroups, each containing each four observations.

Variability within the Subgroups ('Within'):

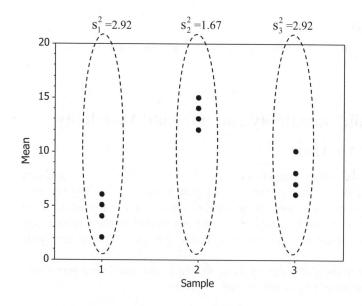

The best estimator of the variance within subgroups is the mean of the subgroup variances. In our case:

$$s_{Within}^2 = \frac{2.92 + 1.67 + 2.92}{3} = 2.50$$

and thus, $s_{Within} = \sqrt{2.50} = 1.58$

Global Variability (Overall)

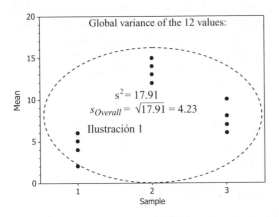

To carry out a capability study of these data, use the arbitrary values of zero and 15 as specifications and in **Estimate** disable the default option (**Use unbiasing constants**). The following results are obtained and the previously computed values can be verified:

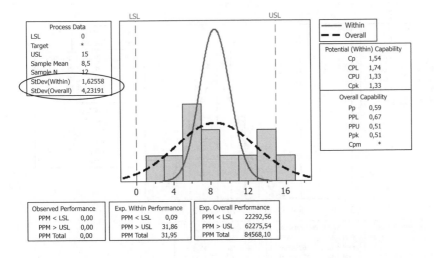

If instead, the option for bias correction factor is selected, the obtained results are slightly different.

14.7 Capability Study when the Sample Size Is Equal to One

14.7.1 Values Taken at Regular Intervals of Time

Example 14.1: A company produces dry food for dogs and uses an extrusion process where the product humidity is a relevant characteristic of quality. To carry out a capability study of the humidity variable, which value must be between 6% and 12%, a one unit sample is taken every 15 minutes and subsequently a measure of the humidity is recorded. Samples are taken during 8 hours obtaining the following 32 humidity measures (entered by rows, order is important):

10.7	9.4	9.4	10.1	10.3	9.8	11.0	10.3
8.5	7.3	10.5	9.2	11.9	11.7	10.9	10.3
11.0	10.6	10.4	11.5	13.5	12.5	12.2	11.8
9.4	12.0	10.4	12.0	13.5	10.4	13.6	11.4

Place the data values in a column named Humidity, and go to:

Stat > Quality Tools > Capability Analysis > Normal, and indicate that the data is contained in the column **Humidity**, that the subgroup size (sample size) is **1**, and that the specification limits are **6** and **12**. The following output is displayed:

Process Capability of Humidity

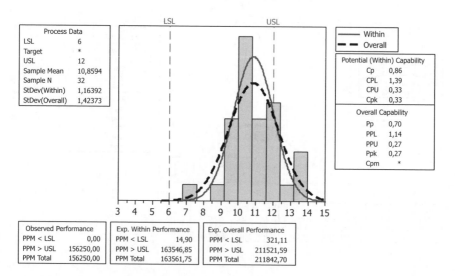

For the sake of clarity, the values of the horizontal scale as well as the width and position of the histogram bars have been changed

In the previous dialog boxes, instead of writing all the values (3 4 5 6 7 8 9 10 11 12 13 14 15), you can write 3:15/1 (that is, values from 3 to 15 with increments of 1).

The process is not capable (Cp<1) and it is clearly not centred. The discrepancy between the within variability (computed in this case from the moving ranges) and the overall variability shows that the process has not remained stable. This can be verified by constructing a plot that shows the temporal evolution of the data (**Graph > Time Series Plot**).

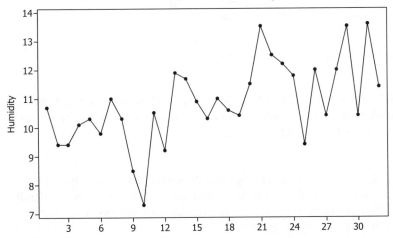

The humidity has an increasing pattern along the day.

14.7.2   Values Taken Consecutively

Example 14.2: The dry food for dogs is packed in bags containing 4 kg. When the packaging process starts, 50 bags are sampled and weighted obtaining the following values:

4.046	3.872	3.942	4.261	3.926	4.085	3.942	4.150	3.943	3.927
4.133	3.995	3.836	4.125	3.862	4.022	3.935	4.157	3.925	4.097
4.128	3.983	4.101	3.861	4.050	4.047	3.951	4.155	4.047	3.898
4.159	3.902	4.037	3.968	3.870	4.103	3.750	4.027	4.093	3.905
3.958	4.113	4.075	3.837	3.978	3.946	4.008	3.897	4.201	3.905

The specifications indicate that the minimum weight must be of 3.8 kg. Carry out a capability study for these data.

Stat > Quality Tools > Capability Analysis > Normal

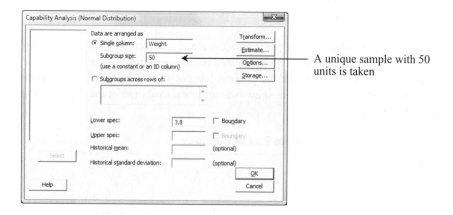

A unique sample with 50 units is taken

In this case, only the within variability is estimated. Consequently, the within and overall standard deviations coincide. Notice that if only one specification limit is available the index Cp cannot be calculated.

Taking 50 units consecutively is not the same as taking 50 units one by one at regular intervals of time. In the first case, only the short-term variability (*Within*) can be estimated. In the second case, the long-term variability (*Overall*) can also be estimated.

Process Capability of Weight

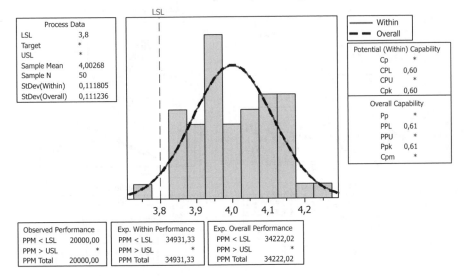

Process Data	
LSL	3,8
Target	*
USL	*
Sample Mean	4,00268
Sample N	50
StDev(Within)	0,111805
StDev(Overall)	0,111236

—— Within
– – Overall

Potential (Within) Capability
Cp	*
CPL	0,60
CPU	*
Cpk	0,60

Overall Capability
Pp	*
PPL	0,61
PPU	*
Ppk	0,61
Cpm	*

Observed Performance		Exp. Within Performance		Exp. Overall Performance	
PPM < LSL	20000,00	PPM < LSL	34931,33	PPM < LSL	34222,02
PPM > USL	*	PPM > USL	*	PPM > USL	*
PPM Total	20000,00	PPM Total	34931,33	PPM Total	34222,02

14.8 A More Detailed Data Analysis (Capability Sixpack)

Consider again the file VITA_C.MTW and go to:

Stat > Quality Tools > Capability Sixpack > Normal

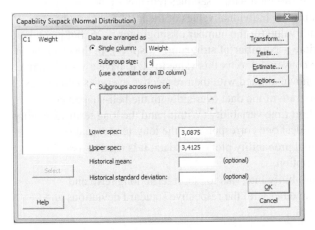

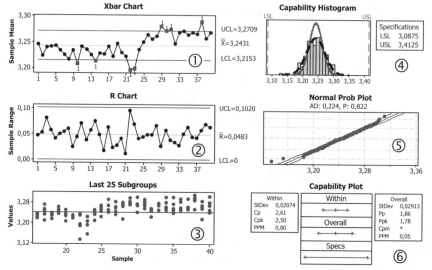

①   Chart representing the data averages in each subgroup. In the example, there is a clear level jump around the sample 22-23, which causes a big discrepancy between the Within and Overall variability measures. The control limits are computed based on the variability within samples (Within).

②   Chart representing the data ranges in each subgroup. In the example, it indicates that the variation within samples remains stable.

③   Chart representing the individual values of the last 25 subgroups. The displayed default value for the number of subgroups is 25, but it can be modified (**Option > Number of subgroups display**). In the example, it would be more convenient to fix this value to 40 to have a visual picture of all the data shown aligned with their respective averages and ranges.

④   Capability histogram of the data overlaid with the bell-shaped curves that show the short time variability (Within) and the long term variability (Overall); the widest one corresponds to the long-term variation.

⑤   Capability normal probability plot of the data. It is used to assess the normality assumption.

⑥   Width of the variability intervals for short term, long term and specifications, together with the respective standard deviations and capability indexes.

15

Capability Studies for Attributes

15.1 File 'BANK'

A bank conducts a study on the satisfaction of its customers. To achieve that, 30 of its agencies are randomly selected, and on each agency, 50 clients are also randomly selected. All of them must undergo an interview. Even though very detailed information is obtained from each interview, at the end, each interview is summarized as either 'Satisfied customer' or 'Unsatisfied customer'. The file BANK.MTW is organized as follows:

Column	Name	Content
C1	Branch	Agency's number
C2	Unsatisfied	Number of customers classified as 'unsatisfied' in that agency

15.2 Capability Study for Variables that Follow a Binomial Distribution

The number of unsatisfied customers in file BANK.MTW is a typical example of a variable that follows a binomial distribution. It must be assumed that all of the agencies have the same instructions, the same products, and that they all follow the same procedures in their relationships with their customers. The question of interest

Industrial Statistics with Minitab, First Edition. Pere Grima Cintas, Lluís Marco-Almagro and Xavier Tort-Martorell Llabrés.
© 2012 John Wiley & Sons, Ltd. Published 2012 by John Wiley & Sons, Ltd.

is to find out which is the percentage of unsatisfied customers that are produced by these products and the standard procedures of the bank.

Stat > Quality Tools > Capability Analysis > Binomial

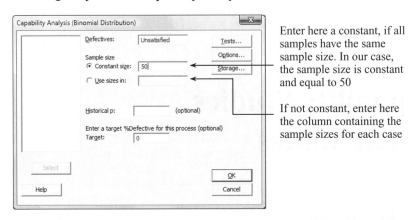

Enter here a constant, if all samples have the same sample size. In our case, the sample size is constant and equal to 50

If not constant, enter here the column containing the sample sizes for each case

Output in the session window:

Description of the type of test not surpassed by some of the points

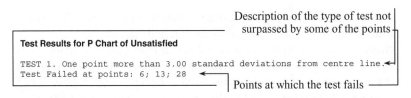

Test Results for P Chart of Unsatisfied

TEST 1. One point more than 3.00 standard deviations from centre line.
Test Failed at points: 6; 13; 28

Points at which the test fails

Graphical window:

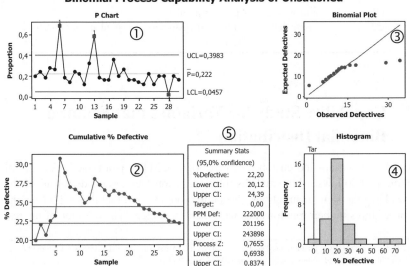

Binomial Process Capability Analysis of Unsatisfied

① Control chart for the proportion of unsatisfied customers. We observe the points outside control as pointed out in the Session window.

② Cumulative proportion of defects. It must stabilize around the mean value to make sure that the number of samples is representative.

③ Representation of the values in binomial probability paper. We observe three points far away from the line.

④ Histogram of the 'defects' together with a reference line that is located on the target value which, in this case, is the default value: 0.

⑤ Summary statistics: A 95% confidence interval for both, the percentage and ppm, of defectives.

Agencies 112 and 212 have a significantly greater number of unsatisfied customers than the other agencies (surely, there are assignable causes). On the other hand, Agency 635 has a number of unsatisfied customers that is even less than normal (we suppose that an assignable cause has been identified too).

 The option **Brush** can be used on the **P Chart** to identify the number of the agencies that appear as anomalies. (For this, one needs to activate **Editor > Set ID Variables** and place C1 in **Variables**.)

Once the assignable causes have been identified, we can repeat the capability study after excluding the three anomalous values (we place an asterisk, '*', upon these values in the worksheet):

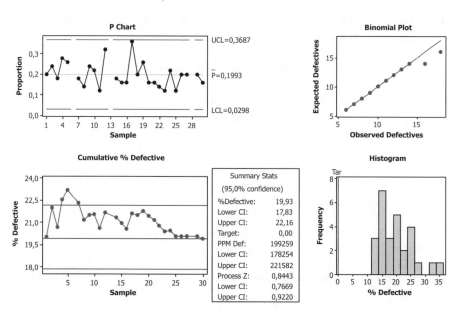

The mean percentage of unsatisfied customers per agency lays in the interval 18–22%, with a confidence level of 95% (in the previous graph: **Summary Stats, %Defective, Lower CI and Upper CI**, rounding the values that appear there). Percentages of unsatisfied customers between 3% and 37% (limits of the P control chart) cannot be considered abnormal.

Another information that could be very valuable is to identify the assignable causes for some agencies behaving very differently from the rest.

15.3   File 'OVEN_PAINTED'

At the end of the painting process of some parts that are dried in an oven, each one of these parts is inspected due to the possible presence of small defects that need to be corrected by hand. The file OVEN_PAINTED.MTW contains the number of defects detected in 40 consecutively inspected parts. The file is organized as follows:

Column	Name	Content
C1	Part	Part number
C2	Num. defects	Number of defects in the part

15.4   Capability Study for Variables that Follow a Poisson Distribution

Here, use the file OVEN_PAINTED.MTW since the number of defects per unit is a typical example of a variable that follows a Poisson distribution.

Stat > Quality Tools > Capability Analysis (Poisson)

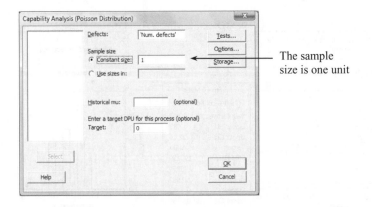

The sample size is one unit

Poisson Capability Analysis of Num. defects

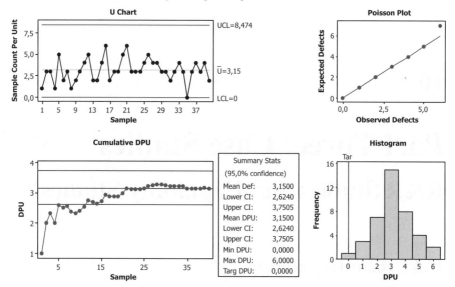

The process remains stable at about 3 defects per unit. The graph of the cumulative average value indicates that the number of samples is sufficient. Additionally, the **Poisson Plot** suggests that the hypothesis that the data follows a Poisson distribution is perfectly reasonable.

16

Part Three: Case Studies
R&R Studies and Capability Studies

16.1 Diameter_measure

Parts are manufactured whose critical dimension is a diameter that has specification limits given by 7.5 ± 0.1 mm. To measure this diameter, a micrometer is used with a resolution of up to thousandths of a millimetre.

The company which produces this part uses two standards on the suitability of the measurement process:

- Standard A: The variability of the measurement system must have a width, measured as 6 standard deviations, of less than 20% of the tolerance interval.

- Standard B: The variability of the measurement system must have a width, defined as 4 standard deviations, of less than 15% of the tolerance interval.

To verify if either of these rules are met and whether the measurement process is working properly, a sample of 10 parts of recent production is taken and 3 operators measure 3 times each of them in randomorder and without knowing which part

Industrial Statistics with Minitab, First Edition. Pere Grima Cintas, Lluís Marco-Almagro and Xavier Tort-Martorell Llabrés.
© 2012 John Wiley & Sons, Ltd. Published 2012 by John Wiley & Sons, Ltd.

they are measuring. The results are stored in file DIAMETER_MEASURE.MTW, in such a way that each row contains the 9 measurements made for each piece:

Column	Content
C1	Number of the part measured
C2-C4	Measures taken by Carlos of each of the parts
C5-C7	Measures taken by Mikel
C8-C10	Measures taken by Pablo

Does the measurement system work correctly? Does it meet any of the two above-mentioned standards?

The data are not organized in an appropriate way so that the study can be carried out. Thus, first proceed to place all measures in a single column by doing the following.

Data > Stack > Columns

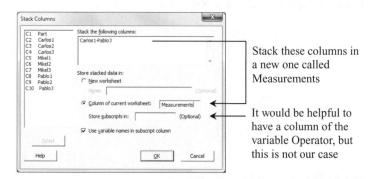

Stack these columns in a new one called Measurements

It would be helpful to have a column of the variable Operator, but this is not our case

To indicate to which operator each measurement belongs, use the following.

Calc > Make Patterned Data > Text Values

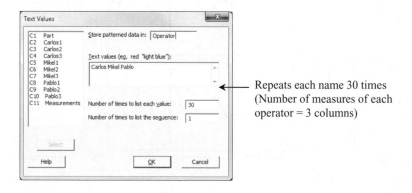

Repeats each name 30 times (Number of measures of each operator = 3 columns)

To create the column with the part number, the option **Make Patterned Data** could be used, but instead it is done via **stack columns**:

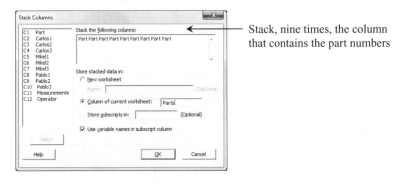

Stack, nine times, the column that contains the part numbers

Once the data are organized as required by Minitab in order to carry out the study, start with the graphical analysis, as follows.

Stat > Quality Tools > Gage Study > Gage Run Chart

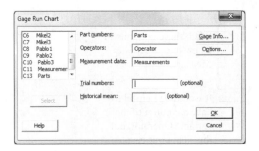

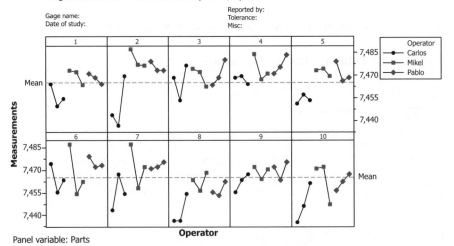

Some conclusions that can be drawn from the previous graph are:

- Pablo's measurements present less variability, being constant along the time.

- Carlos' measurements show the highest variability.

- Among Mikel's measurements, the variability tends to increase (parts 6, 7 and 10).

- Carlos' measurement values are, in most cases, the lowest.

Before carrying out the analytical study, observe that we are dealing with a typical crossed design. Hence, in order to continue the study, do as follows.

Stat > Quality Tools > Gage Study > Gage R&R Study (Crossed)

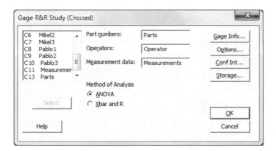

And in **Options**:

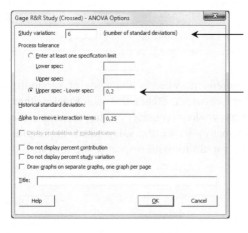

Amplitude of variation which is equivalent to 99.7 % of all values. This is the default value, which helps us to judge the compliance of the standard A

Width of the tolerance interval (±0.1)

Two-Way ANOVA Table With Interaction

Source	DF	SS	MS	F	P
Parts	9	0.0025457	0.0002829	4.6166	0.003
Operator	2	0.0037278	0.0018639	30.4209	0.000
Parts*Operator	18	0.0011029	0.0000613	0.7933	0.700
Repeatability	60	0.0046340	0.0000772		
Total	89	0.0120104			

There are differences among both, parts and operators (very small p-values). There is no part-operator interaction (p >0.25).

Alpha to remove interaction term = 0.25

Two-Way ANOVA Table Without Interaction

Source	DF	SS	MS	F	P
Parts	9	0.0025457	0.0002829	3.8458	0.000
Operator	2	0.0037278	0.0018639	25.3421	0.000
Repeatability	78	0.0057369	0.0000735		
Total	89	0.0120104			

Since the interaction is not significant, a second ANOVA table excluding interaction is shown

Gage R&R

Source	VarComp	%Contribution (of VarComp)
Total Gage R&R	0.0001332	85.14
Repeatability	0.0000735	47.00
Reproducibility	0.0000597	38.14
Operator	0.0000597	38.14
Part-To-Part	0.0000233	14.86
Total Variation	0.0001565	100.00

... ...

Source	StdDev (SD)	Study Var (6 * SD)	%Study Var (%SV)	%Tolerance (SV/Toler)
Total Gage R&R	0.0115424	0.0692546	92.27	34.63 ←
Repeatability	0.0085761	0.0514566	68.56	25.73
Reproducibility	0.0077252	0.0463511	61.76	23.18
Operator	0.0077252	0.0463511	61.76	23.18
Part-To-Part	0.0048225	0.0289351	38.55	14.47
Total Variation	0.0125094	0.0750563	100.00	37.53

Standard A is not met, as the measurement system's amplitude of variability (99.7%) is greater than 20% of the tolerance width. Standard B is not met neither, since, substituting 6 by 4 in the **Options** window (considering that the amplitude of the bell curve is 4σ, that is, approximately 95%), the value which appears as percentage of the measurement system's variability with respect to the tolerance is 23.08% (> 15%).

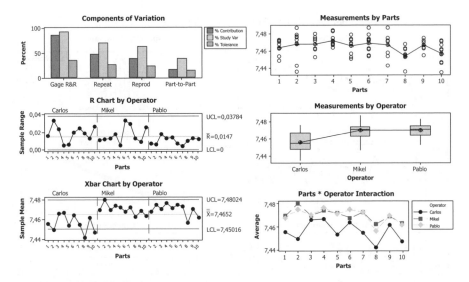

An evident conclusion, already noticed in the initial graphical analysis, is that Carlos' values lie clearly below those of both other operators. This indicates that the operators should be trained to improve the measurement system.

16.2   Diameter_capability_1

The previously mentioned part must be produced for a client that requires that the capability index of the critical diameter meets $Cp > 1.33$. A first series is produced during two days in order to verify that all of the client's requirements are met.

   To carry out the capability study, five parts are taken every half an hour (16 samples each day, 32 samples in total), and the results obtained are stored in file DIAMETER_CAPABILITY_1.MTW. The data are organized in rows, that is, the first row contains the values of the first sample, the second row the values of the second sample, etc.

   Is the requirement $Cp > 1.33$ met?

Use: **Stat > Quality Tools > Capability Sixpack > Normal**

   Since the data are organized by rows, use the option **Subgroups across rows of**: and introduce C2-C6. Remember that the specifications are 7.5 ± 0.1.

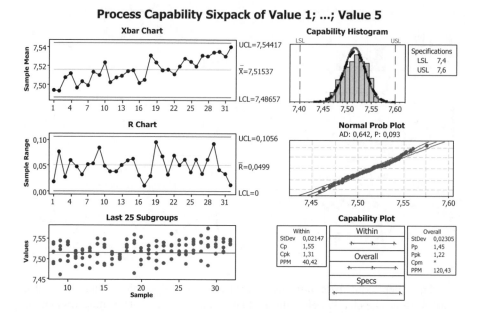

An evident conclusion obtained from the graph of the sample means (**Xbar Chart**) is that the process becomes decentred towards larger values. The reason for that should be studied; perhaps it is a badly calibrated machine, a wear and tear of the tools, etc. Concerning the range chart, there is nothing to object. The histogram is decentred with respect to the target value; the default values of the scale are changed to show this more clearly.

The required capability index Cp > 1.33 is met in this case (also Pp > 1.33), although it appears that this requirement would not be met in case of large runs without adjusting the machine. If the machine is adjusted daily, the index Cp will be larger.

16.3   Diameter_capability_2

During the series production of the previous part, each day, after adjusting the machine, 50 parts are consecutively produced and later a capability study is carried out with them to assure that the adjustment was done correctly.

The 50 values used for the capability study on a particular day are stored in the file DIAMETER_CAPABILITY_2.MTW. Is everything correct?

Proceed to carry out a capability study going to:
   Stat > Quality Tools > Capability Analysis > Normal

In the dialog box, it is only necessary to indicate that the data are in the **Single column**: C1, that the **Subgroup size** is 50, and that the specification limits are 7.4 and 7.6.

Process Capability of 50 values

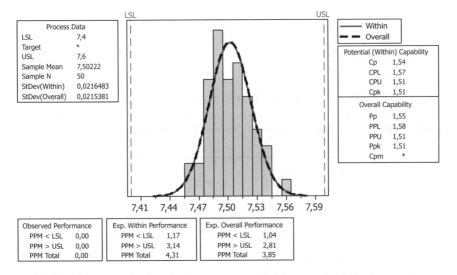

Observed Performance		Exp. Within Performance		Exp. Overall Performance	
PPM < LSL	0,00	PPM < LSL	1,17	PPM < LSL	1,04
PPM > USL	0,00	PPM > USL	3,14	PPM > USL	2,81
PPM Total	0,00	PPM Total	4,31	PPM Total	3,85

Everything appears correct in the graphical output of the capability study. Also, it is important to analyze whether there is any tendency among the data. In this case, we cannot use **Capability Sixpack** since only one single point is shown on the control graph (there is only one sample of size 50). Thus, it is more appropriate to make a control graph for the individual observations, as shown below.

Stat > Control Charts > Variables Charts for Individuals > Individuals

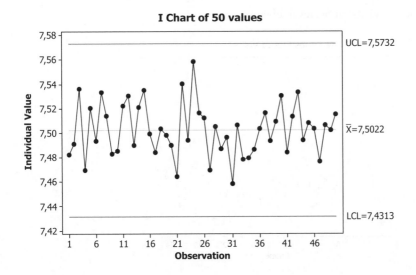

In the above graph, nothing in particular is observed. Hence, the production can be started.

16.4   Web_visits

The file WEB_VISITS.MTW contains the number of daily visits during October and November 2010 to a website which provides information about training activities in the area of quality control. The file is organized as follows:

Column	Name	Content
C1	Date	Date
C2	Day of the week	Weekday
C3	Visits	Number of visits

Based on these data, a capability study of the number of daily visits is carried out with the objective to assess, later on, the impact of the actions for promotion which are intended to be performed.

An important piece of information is that on 14 October a press announcement was released on the institution which organizes these activities. Therein, it was indicated that more information could be found on that web page. During the second week of November, from day 11, a booklet was distributed within which it was also said that more information could be found on that web page.

Start carrying out an exploratory data analysis using a time series plot obtained by:
   Graph > Time Series Plot: Simple

Simply place 'Visits' in **Series** to obtain:

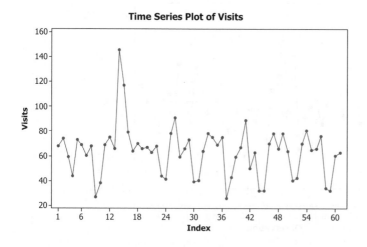

A large peak is observed around observation 14 as well as a certain periodicity within the data. The option **Brush** allows us to identify which day each point corresponds with (**Set ID variables**: Date; Day of the week was also used).

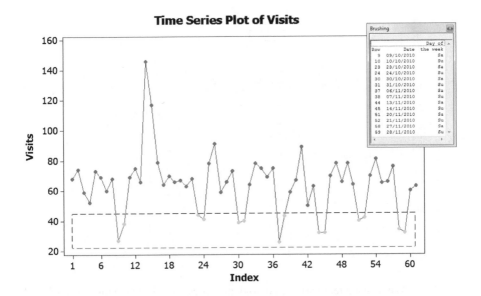

All valley points correspond to weekends (Saturdays and Sundays). The peak around observation 14 corresponds to Thursday, 14 October, the day of the advertisement in the press. This peak effect decays, but lasts several days so that the following weekend, the number of visits to the web page do not fall as on other weekends. Additionally, the distribution of the booklet does not seem to have any effect on the number of visits to the web page.

Since there are two types of days, workdays and weekends, proceed to carry out a capability study only for the workdays, discarding also days affected by the peak on 14 October.

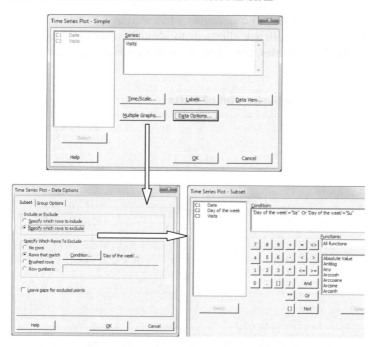

 When using text variables in logical expressions, its value must be enclosed in double quotation marks.

Without the weekends, after choosing the option that allows leaving the space corresponding to the points excluded, the graph looks as follows:

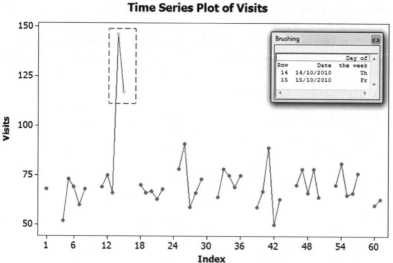

Notice that only the days of the press announcement stand out (Thursday and Friday) and since the causes for these are clear (special causes), eliminate both values from the study replacing them with an asterisk.

Next, to carry out the capability study, create a new column that contains only the values of interest for the study. Do this through **Data > Copy > Columns to columns** and then click on button **Subset the Data**, which allows a selective copy.

Since the number of daily visits can be treated as a variable which follows a Poisson distribution, do the following.

Stat > Quality Tools > Capability Analysis > Poisson

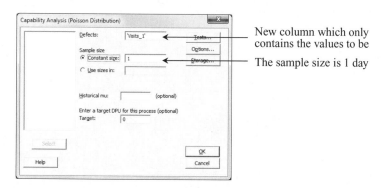

New column which only contains the values to be

The sample size is 1 day

There are no points out of control, the values follow a Poisson distribution reasonably well, and the number of visits remains fairly stable around its mean value. It seems that the data are correct to draw conclusions.

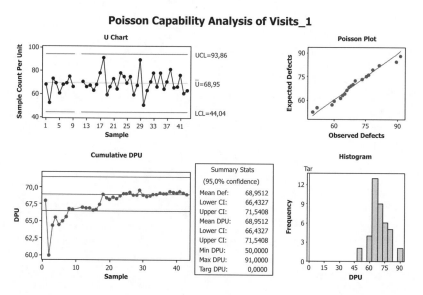

Poisson Capability Analysis of Visits_1

The average number of daily visits on workdays is around 70. Number of visits below 45 and above 95 (using round numbers) can be attributed to special causes.

We have assumed that the distribution of the booklet had no impact on the number of visits. Another possible interpretation could be that the number of visits decreases, and that the increase due to the booklet makes this trend unnoticeable. To clarify this question, it is necessary to wait and see how the data evolve in December.

Part Four

MULTI-VARI CHARTS AND STATISTICAL PROCESS CONTROL

Multi-vari charts are not part of the classic tools for data analysis but they are conceptually very simple and in many cases they are the best way to identify the different sources of process variability. Naturally, and once again, for this to work, data should have been collected in an appropriate manner. Multi-vari charts are usually considered an exploratory data analysis technique, but we have preferred to place it in Part 4, along with control charts, because it shares with them the objective of fighting against variability.

Statistical Process Control (SPC) is one of the best known and more widely used quality tools. Its aim is to keep the process working at its 'normal state' – with only random variability causes affecting it – by releasing alarm signals when something strange – a special cause of variability – is disturbing the process. In this sense they are usually considered a control tool, but it is clear that they are also – and perhaps mainly – an improvement tool, as the opportunity is given to learn from the process, by showing what causes affect it and in which way; thus, providing the information needed to improve the process by incorporating the good and eliminating the harmful ones.

The generic name of SPC is an umbrella for several types of control charts that all work with the same general principle stated above. Control charts are classified into two big groups depending on whether they are aimed at controlling quantitative variables or attributes (features that the product has or has not). It is easy to identify if an attribute or a variable chart is needed but sometimes is not that easy to choose the right graph in each category.

As a general rule, when controlling variables it is always more efficient to use charts representing averages than individual observations. The idea is to take, at regular intervals, a small sample of homogeneous items – many times the homogeneity criterion is fulfilled by taking items produced consecutively – and compute and represent the average in the chart. Apart from being very efficient to detect changes in the process mean, this allows us to monitor the process variability by also representing the range of the sample on a complementary chart. The two charts are represented together and called $\bar{X} - R$ charts. Sometimes the standard deviation of the small sample is used to produce an $\bar{X} - S$ chart.

Sometimes it is impossible or does not make sense to compute averages, for example when controlling the temperature of a process. In those cases individual observations charts are used; they are usually complemented with a variability chart based on the calculation of mobile ranges. Minitab also allows the use of slightly more sophisticated charts that give different weight to the observations.

The following table summarizes the types of attributes charts, depending on whether the binomial or Poisson distribution is appropriate to model the variable to control. From the onset, it may seem difficult to identify the proper distribution, but it is not. The table provides hints that make it very easy:

Distribution	Charts	Characteristics
Binomial	P, NP	– You can count occurrences vs. nonoccurrences (good parts / bad parts, orders delivered on time against orders not delivered on time, etc.). – There is a limit of occurrences (maximum number of defective parts will be the total number; at most all orders received will be delivered late, etc.).
Poisson	C, U	– It is not possible to count nonocurrences (it is possible to count the number of visits to a website, but not the number of 'nonvisits', or you can count the number of phone calls to a switchboard for an hour, but not the number of 'no calls'). – There is no limit (at least from a theoretical point of view) to the number of occurrences. There is no theoretical limit to the number of visits to a web page or the number of calls to the switchboard.

When the binomial distribution applies and the sample size is constant (for example, defects are counted in samples of 50 units) both: P (ratio of defects) and NP (number of defects) charts can be used. But if the sample size is not constant, for example if you look at production defects per hour and production per hour is not constant, then only the P chart can be used. With respect to the Poisson distribution when the sample size is constant (annual number of failures in an elevator), a C chart is the appropriate one, but if the size is variable, as when controlling the number of coffees consumed in a vending machine between two replacement visits when the time between visits is not constant, a U chart is the one to use.

When exercises or case studies on this topic are presented, the data for the entire period analyzed is available and what is done is an a posteriori analysis. This is not a very realistic situation. We should not forget that the power of control charts relies on representing the points on the chart as the data is gathered from the process. The chart will then show if something 'strange' is happening, if there are signs that something is starting to go wrong or, on the contrary, if everything is 'normal'. This is the way to improve the process: look for the causes of out of control signals and react quickly, and leave it alone if no special causes are acting.

17

Multi-Vari Charts

17.1 File 'MUFFIN'

A local bakery decides to commercialize their famous raspberries muffins through a chain of supermarkets. This requires industrializing the baking process. Their first attempts are unsuccessful due to excessive variability in the height of the muffins. A data collection is organized to try to find the main source of variability. Every hour, three consecutive trays (each having four muffins) are taken from the process, and the height of the four muffins is measured.

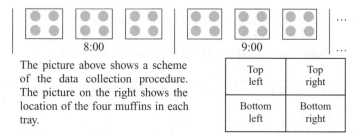

The picture above shows a scheme of the data collection procedure. The picture on the right shows the location of the four muffins in each tray.

| Top left | Top right |
| Bottom left | Bottom right |

File MUFFIN.MTW contains the data, and is organized as follows:

Column	Name	Contents
C1	Time	Time when the sample was taken
C2	Tray	Number of tray
C3	Position	Muffin location on the tray (top left, bottom left, top right, bottom right).
C4	Height	Muffin height

Industrial Statistics with Minitab, First Edition. Pere Grima Cintas, Lluís Marco-Almagro and Xavier Tort-Martorell Llabrés.
© 2012 John Wiley & Sons, Ltd. Published 2012 by John Wiley & Sons, Ltd.

17.2   Multi-Vari Chart with Three Sources of Variation

Data in MUFFIN.MTW file were collected taking into account that height variability could be due to:

- differences between the left and right parts of a tray;
- differences between top and bottom parts of a tray;
- variability among muffins baked in consecutive trays (short term variability);
- variability due to time (long term).

Multi-vari charts allow visualizing the relative importance of each source of variability. To achieve this, do as follows.

Stat > Quality Tools > Multi-Vari Chart

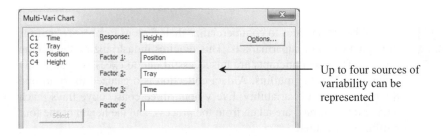

Up to four sources of variability can be represented

With all options by default, the following chart is obtained:

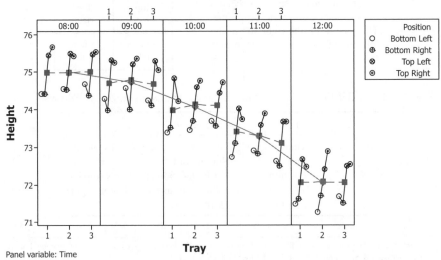

The most important source of variability is due to changes over time (probably, because temperature in the oven tends to decrease, preventing muffins to acquire their nominal height). It is represented by the line connecting diamond-shaped points. The second source of variability is due to the position of the muffin in the tray. Finally, the last source of variability is short-term variability among trays. This source of variability is much smaller than the others.

Multi-vari options allow removing the connecting lines. This is sometimes very useful to make the graph more clear to interpret. For instance, to remove the lines connecting positions in each tray, go to **Options**:

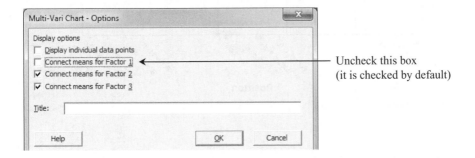

Uncheck this box (it is checked by default)

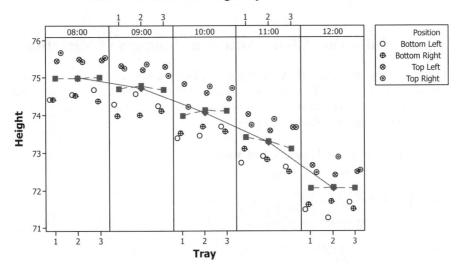

Panel variable: Time

The appearance of the chart depends on the order in which the sources of variation are introduced in the dialog box. For example, if the order is: Tray – Position – Time, the following graph (probably easier to interpret than the previous one) is obtained:

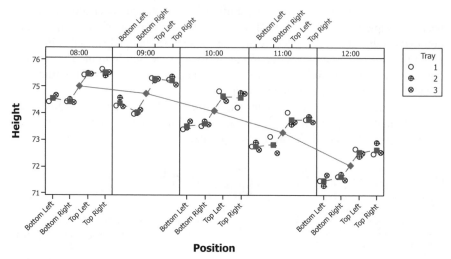

Panel variable: Time

 Multi-vari charts' appearance depends upon the order in which each factor is introduced. Since some charts' appearances are more clear to interpret than others, the order of several factors should be tested.

17.3   Multi-Vari Chart with Four Sources of Variation

Continue using data from file MUFFIN.MTW. Break down column 'Position' into two different columns: 'HorizontalPos' (for horizontal position, left or right) and 'VerticalPos' (for vertical position, top or bottom). For instance, create column 'HorizontalPos' in the following manner.

Calc > Make Patterned Data > Text Values

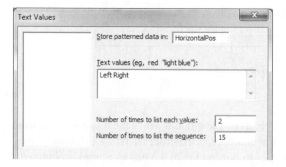

To create column 'VerticalPos', the procedure is the same, but now the text values are Top and Bottom, the number of times to list each value: 1, and the number of times to list the sequence: 30.

With both new columns created, do:

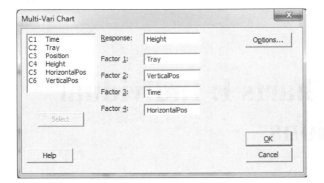

The result is the following:

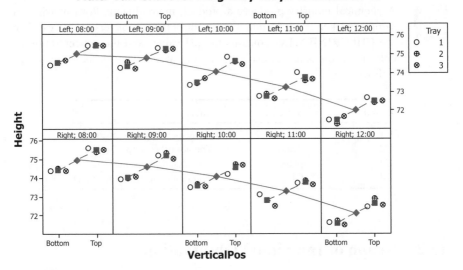

Panel variable: HorizontalPos, Time

Take into account that the last factor in the multi-vari chart dialog box is the one used to divide the graph in the top and bottom parts. The third factor is the one placed in the horizontal axis; thus the best option is setting this factor as the one showing the temporal sequence

In general, multi-vari charts are better interpreted if the factor that shows the temporal evolution is in the third place.

18

Control Charts I: Individual Observations

18.1 File 'CHLORINE'

 A chemical industry produces a product, in continuous flow, of which a sample is taken every 15 minutes to control the parameters of interest. The file CHLORINE.MTW contains the pH values and the concentration of chlorine in the samples taken during one week. The file is organized as follows:

Column	Variable	Label
C1	Date	Corresponding date of the data
C2	Time	The hour the sample is taken
C3	pH	pH value of the sample
C4	Cl	Chlorine concentration (mg/l)

18.2 Graph of Individual Observations

Create a graph for the 32 samples of the pH values (file CHLORINE.MTW) obtained in the last day. To achieve that, first, separate the values which correspond to that last day (which is a Friday; that is why we add an 'F' at the end of the variable names in the new columns). This is done by copying the selected columns, as follows.

Industrial Statistics with Minitab, First Edition. Pere Grima Cintas, Lluís Marco-Almagro and Xavier Tort-Martorell Llabrés.
© 2012 John Wiley & Sons, Ltd. Published 2012 by John Wiley & Sons, Ltd.

Data > Copy > Columns to Columns

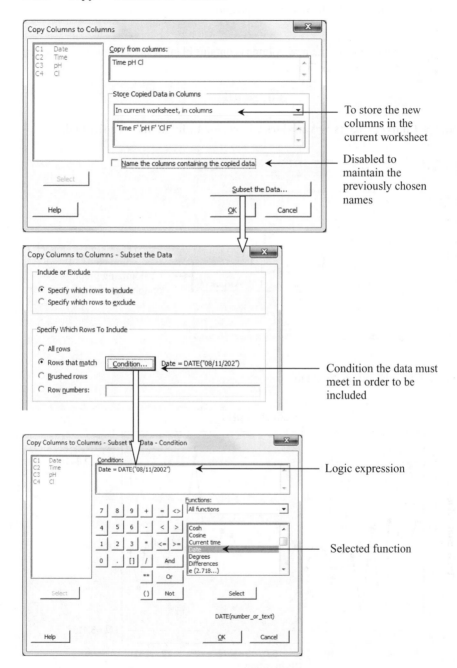

Remember: Just write the names of the new columns and Minitab will create them in the first empty position.

If the name of the new columns contains blank spaces, it must be enclosed in single quotes: 'Time F'.

Stat > Control Charts > Variables Charts for Individuals > Individuals

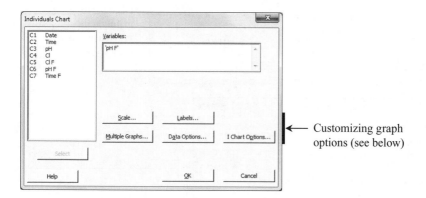

Customizing graph options (see below)

Use all default options to obtain:

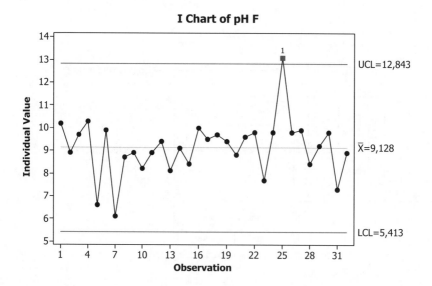

I Chart of pH F

18.3   Customizing the Graph

Scale

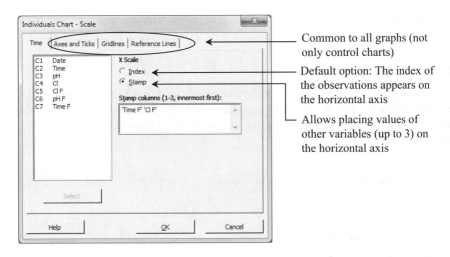

Common to all graphs (not only control charts)

Default option: The index of the observations appears on the horizontal axis

Allows placing values of other variables (up to 3) on the horizontal axis

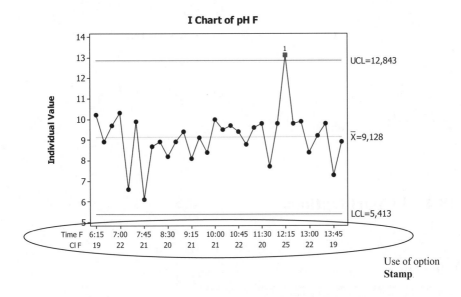

Use of option **Stamp**

Remember: By double-clicking on any axis its values can be changed.

Labels

This allows the addition of titles and comments at the bottom.

Multiple Graphs

If multiple graphs are drawn, the vertical scale can be forced to be same for all.

Data Options

This allows the selection of the values to be represented. For example, with option **Brush (Editor > Brush,** with the graph window being active) we can see that the point outside the control limits is the one in row 25. If you want to exclude it, do the following:

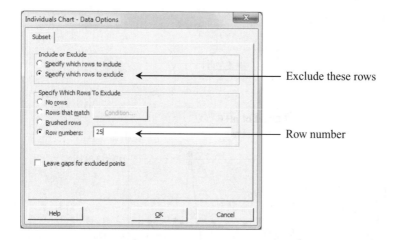

18.4   I Chart Options

Parameters

Allows entering the mean and standard deviation of the process and Minitab calculates the limits with both values. By default, these parameters are estimated from the data.

Estimate

It allows you to skip samples (subgroups) from the calculation of the limits. This is useful when you have the certainty of the presence of special causes in a sample.

S Limits

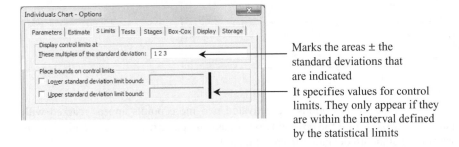

Marks the areas ± the standard deviations that are indicated

It specifies values for control limits. They only appear if they are within the interval defined by the statistical limits

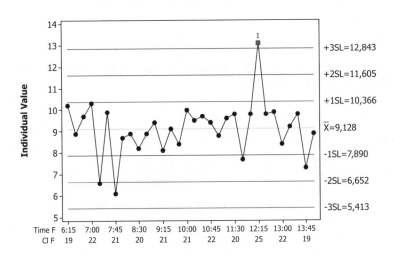

Tests

By default, only the symptom of special cause for one point beyond the limits (pattern 1) is activated. There are eight symptoms available and it is possible to activate the ones of interest by this option.

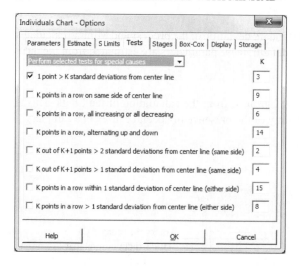

In our case, if all options are activated two more points appear marked with pattern 5 (two out of three beyond two sigmas) and pattern 7 (15 consecutive points within one standard deviation from the mean).

 Activating all alarm signs increases the risk of false alarms.

Stages
Mark groups on the graph. We go back and change variable 'pH F' (Friday's pH value) by 'pH' (all days' pH values).

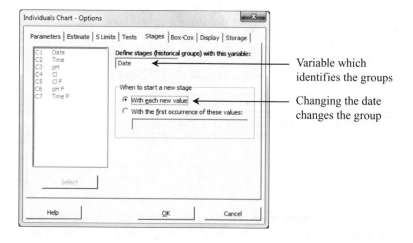

I Chart of pH by Date

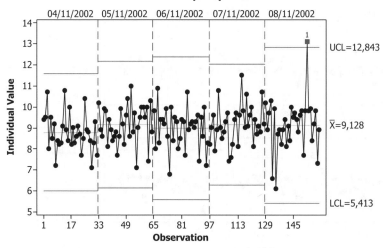

Note: All signals indicating that the process is out of control are still enabled.

Box-Cox

This transforms the original data using the Box-Cox transformation; it is used to normalize the data.

Display

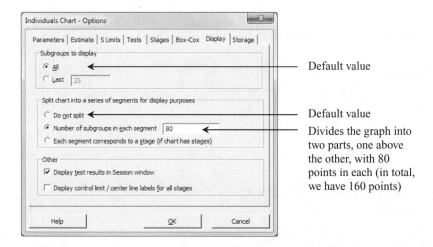

Default value

Default value

Divides the graph into two parts, one above the other, with 80 points in each (in total, we have 160 points)

I Chart of pH by Date

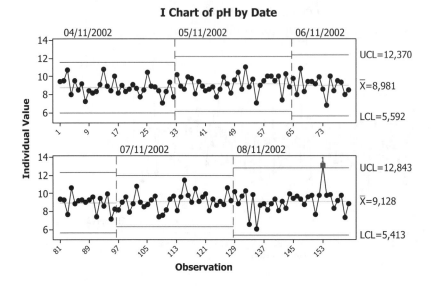

Storage

This stores the indicated values in the worksheet.

18.5   Graphs of Moving Ranges

Consider again the file CLORINE.MTW and draw the graph of the moving ranges of the pH values on Friday going to the following.

Stat > Control Charts > Variables Charts for Individuals > Moving Range

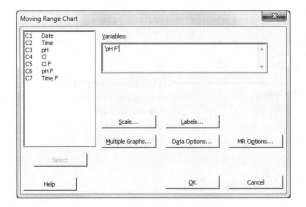

The options are basically the same as in the case of individual observations.

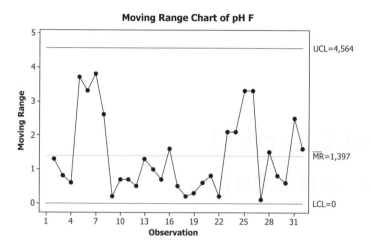

18.6   Graph of Individual Observations – Moving Ranges

These are the graphs we have seen already, but now they are plotted together in the same graphical window.

Stat > Control Charts > Variables Charts for Individuals > I-MR

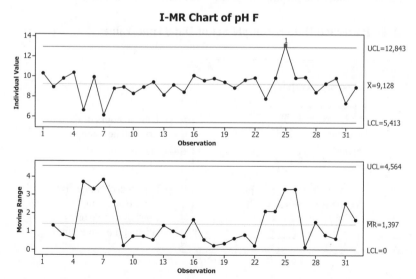

19

Control Charts II: Means and Ranges

19.1　File 'VITA_C'

Consider again the file VITA_C.MTW presented in Chapter 14 (capability studies). From a machine producing pills, samples of 5 units are taken every 15 minutes during 10 hours. Column C1, the only one containing data, contains measures corresponding to the weights of the sampled pills (40 samples × 5 pills/sample = 200 pills).

To illustrate better the possibilities of these type of plots, we create two new columns: the first contains the time (in hours) in which the datum is taken and the other contains a number identifying the worker operating the machine.

Calc > Make Patterned Data > Simple Set of Date/Time Values

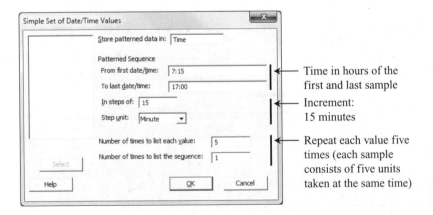

Time in hours of the first and last sample

Increment: 15 minutes

Repeat each value five times (each sample consists of five units taken at the same time)

Industrial Statistics with Minitab, First Edition. Pere Grima Cintas, Lluís Marco-Almagro and Xavier Tort-Martorell Llabrés.
© 2012 John Wiley & Sons, Ltd. Published 2012 by John Wiley & Sons, Ltd.

With respect to the worker, assume that the first 25 samples are taken from the machine operated by worker A, and the other 15 from the machine operated by worker B. This requires us to introduce in column C3 125 (25 × 5) A's and 75 (15 × 5) B's.

Since the number of A's is distinct from the number of B's, the option **Make Patterned Data** cannot be used (although it could be used and then erase the last 50 B's). A quick way to do it is to write directly in the session window (make sure the option **Editor** > **Enable commands** is checked):

MTB > set c3 ◄──── The prompt MTB> changes in the next line after typing set c3

DATA> 125(1) ◄──── The first number indicates the number of replicates and the

DATA> 75(2)   second, between parenthesis, the value to be repeated

DATA> end ◄──── Once all values are introduced, type end and the original

MTB >   prompt MTB> appears again

The name of the column must be introduced 'by hand'.

19.2   Means Chart

Stat > **Control Charts** > **Variables Charts for Subgroups** > **Xbar**

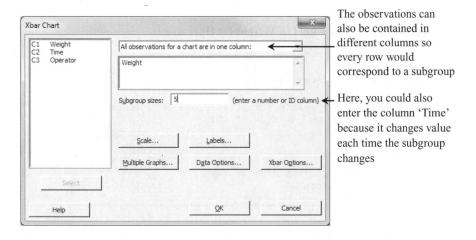

The observations can also be contained in different columns so every row would correspond to a subgroup

Here, you could also enter the column 'Time' because it changes value each time the subgroup changes

The options are identical to the ones used in charts for individual observations (see Chapter 18). We will use them in the following way:

Scale > **Time:** Mark the option **Stamp** and place as variable 'Time'.

Xbar Options > **Tests:** Mark the 8 tests (or choose the option **Perform all tests for special causes**).

Xbar Options > **Stages: Define stages**. . . Operator.

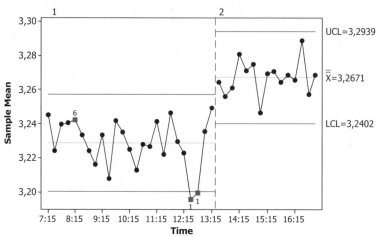

In the session window one finds information on the points out of control. The chart indicates only one test for each point; even if, as happens with point 23, it fails in several.

Test Results for Xbar Chart of Weight by Operator

```
TEST 1. One point more than 3.00 standard deviations from
centre line.
Test Failed at points: 22; 23

TEST 5. 2 out of 3 points more than 2 standard deviations
from centre line (on one side of CL).
Test Failed at points: 23

TEST 6. 4 out of 5 points more than 1 standard deviation
from centre line (on one side of CL).
Test Failed at points: 5
```

19.3   Graphs of Ranges and Standard Deviations

These charts are constructed in the same way as the graphs of means:

Stat > Control Charts > Variables Charts for Subgroups > R
Stat > Control Charts > Variables Charts for Subgroups > S

19.4   Graphs of Means-Ranges

Stat > Control Charts > Variables Charts for Subgroups > Xbar-R
Choosing the options already used for the Xbar graph, one obtains:

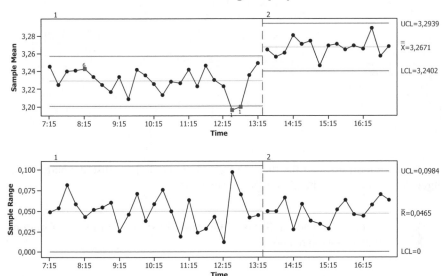

Xbar-R Chart of Weight by Operator

The examples of control charts created on a set of data can give the false impression that the data are first taken and then, once all are collected, the graph is drawn. This is not the case. To rapidly react to the coming problems, the chart must be created as the data are being collected, point by point.

19.5   Some Ideas on How to Use Minitab as a Simulator of Processes for Didactic Reasons

Using the possibility of automatically updating the graphs and the use of macros or executable files, the evolution of a control chart can be simulated. The next file should be first created using: **Tools > Notepad**.

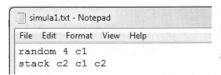

Save the file in the desktop with the name 'simula1' (the extension .txt is added automatically)

To indicate Minitab where to find the files to be executed later, do: **Tools** > **Options:**

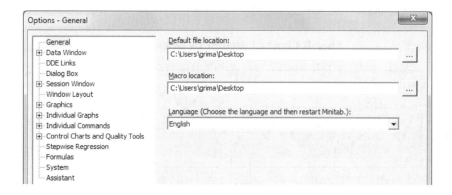

We have indicated that the files are located on the desktop (use the button with three dots to locate the situation in your computer).

 The changes made in **Tools** > **Options** are stored and remain from one Minitab session to another.

Now, place four values from a Normal (0; 1) in column C2. This can be done via **Calc** > **Random Data** > **Normal**, but it can also be done simply by typing the following command in the session window (the option **Enable Commands** must first be activated):

```
MTB > random 4 c2
```

Create now the chart Xbar-R using: **Stat** > **Control Charts** > **Variables Charts for Subgroups** > **Xbar-R**

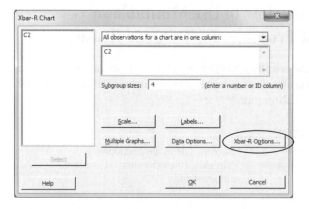

To avoid that limits changing as the points appear, use the option:

Xbar-R Options > **Parameters**: **Mean**: 0; **Standard deviation**: 1

And hit the button **OK** so the Xbar and R charts appear, both with only one point. Double-clicking on the scale, the minimum and maximum values of the horizontal axes of the two plots are modified.

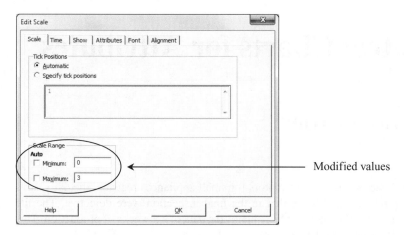

Modified values

Finally, right-click on any point on the plot and activate the option: **Update Graph Automatically.**

Now, execute it 29 times using: **File** > **Other files** > **Run an Exec**

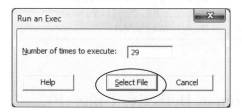

Choose **Select File** and in the text box for the name type *.txt so the name of the file appears in the upper window. Double-click on the name and observe how the graph evolves (make it visible if it is hidden: **Window** > **Xbar-R Chart of C2**).

20

Control Charts for Attributes

20.1 File 'MOTORS'

A factory that makes motors for small appliances registers the daily units produced, and those that in the final inspection were defective. The file named MOTORS.MTW contains the data of six weeks of production. This file is organized as follows:

Column	Variable	Label
C1	Date	Date corresponding to data in the row.
C2	Production	Daily number of produced motors.
C3	Number of defects	Daily number of defective motors.

20.2 Plotting the Proportion of Defective Units (P)

Using the data in file MOTORS.MTW, proceed to construct a P chart as follows.

Industrial Statistics with Minitab, First Edition. Pere Grima Cintas, Lluís Marco-Almagro and Xavier Tort-Martorell Llabrés.
© 2012 John Wiley & Sons, Ltd. Published 2012 by John Wiley & Sons, Ltd.

Stat > Control Charts > Attributes Charts > P

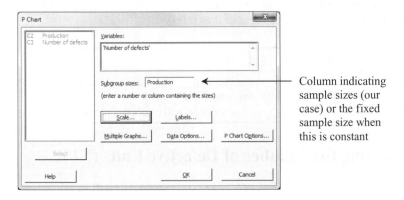

Column indicating sample sizes (our case) or the fixed sample size when this is constant

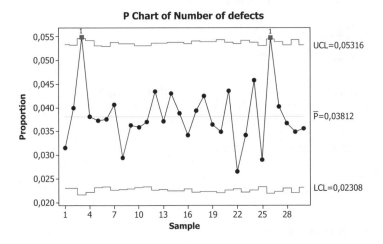

The control limits depend on the sample sizes. When the sample sizes vary, the limits also vary.

Since the number of daily motors produced is very similar, we can choose a constant subgroup size in **subgroup sizes**. In this case 1350 (approximately the average daily production) and consequently a chart with constant control limits would be obtained.

20.3   File 'CATHETER'

A factory that makes catheters for hospital use carries out an inspection of the welding process after the most rigid part is welded to the tip, which is softer. Every hour, a batch of 100 catheters is inspected to detect the possible presence of traces. If the welds have traces, the catheter is defective.

The file named CATHETER.MTW contains the data corresponding to the number of defective catheters, in a batch of 100, during the last week of production. This file is organized as follows:

	Variable	Label
C1	Date	Date corresponding to data in the row
C2	Time	Inspection time
C3	Number of defects	Number of defective catheters per batch

20.4   Plotting the Number of Defective Units (NP)

Using data in file CATHETER.MTW, proceed to construct a NP chart as follows.

Stat > Control Charts > Attributes Charts > NP

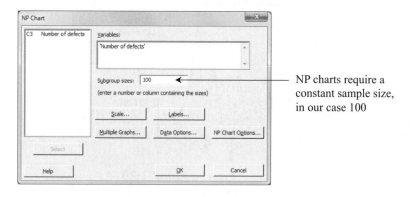

NP charts require a constant sample size, in our case 100

Using the default options one obtains:

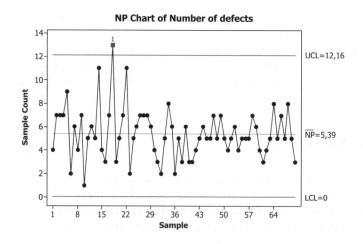

One point lies above the upper control limit. A special cause was found for this unusual increase in the number of defective catheters (a defective lot of raw material), making it reasonable to recalculate the control limits after removing such a singular point.

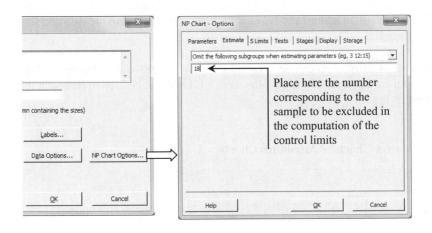

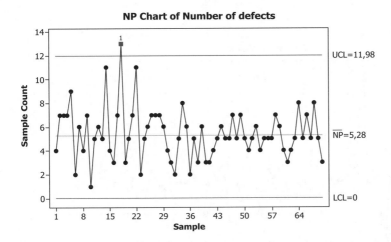

After the recalculation, as expected, the upper control limit is now slightly smaller.

 Although Minitab allows the use of variable sample sizes when constructing NP charts, that is not appropriate. The best practice to facilitate NP charts interpretation is to use a constant sample size, whenever possible. If a constant sample size cannot be used, it is better to draw a P chart instead.

20.5   Plotting the Number of Defects per Constant Unit of Measurement (C)

Consider the file VISITS_WEB.MTW already described in Chapter 16 (Capability analysis). This file contains the number of visits to a website during October and November 2010, specifying also the date and the weekday when those visits were produced.

The number of daily visits can be assumed to follow a Poisson distribution, and as the unit of measurement is constant (one day), a C chart can be used to represent its evolution.

Stat > Control Charts > Attributes Charts > C

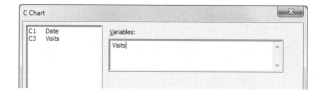

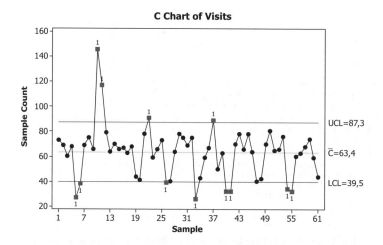

The marked peak observed around day 10 is due to a newspaper advertisement of this website, clearly implying the presence of a special cause. Additionally, the so-called valleys, many of which also lie outside the limits, correspond to weekends when a remarkably smaller number of visits are produced in contrast to working days.

Using the **Data options** button, we remove the data corresponding to Saturdays and Sundays from the chart:

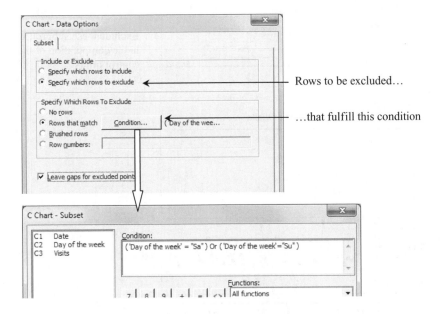

Then, the following chart is obtained, where the limits were recomputed based only on the remaining points.

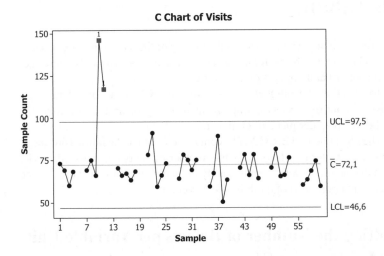

Use the **C Chart Options** button to also exclude from the limits computation the values where the marked peak occurs (at observations 10 and 11). These observations can be identified by using Editor > Brush and clicking on them.

C Chart Options > Estimate

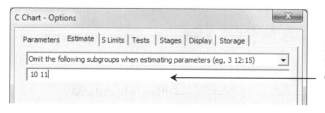

Exclude these subgroups
in the computation of the
control limits

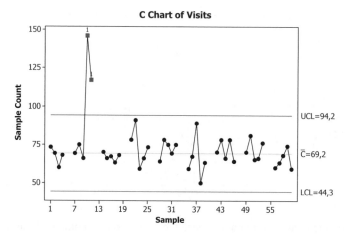

20.6   File 'FABRIC'

A textile company wants to assess the quality of the dyeing process used in the production of fabrics. It has been observed that occasionally a stronger small spot of paint appears on the fabric.

To check if the number of spots remains within a normal range, a statistical control process is implemented. Every quarter of an hour, the number of spots per piece of cloth is recorded.

The file named FABRIC.MTW contains the data of the fabrics produced in one day. Column C1 contains the number of spots per piece of fabric. Since the pieces of cloth are not exactly the same, in column C2 the surface area (in m^2) per piece of cloth is also recorded.

20.7   Plotting the Number of Defects per Variable Unit of Measurement (U)

Similarly to previous cases, using the data in file FABRIC.MTW, proceed to draw a U chart as follows.

Stat > Control Charts > Attributes Charts > U

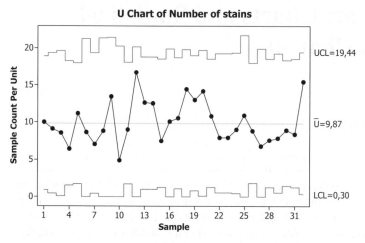

Using all default options, one obtains:

Tests performed with unequal sample sizes

The process is under statistical control. The limits vary because the sample sizes also vary.

 When the control limits are not constant, the values of the upper and lower limits shown by Minitab are the ones from the last sample.

21

Part Four: Case Studies

Multi-Vari Charts and Statistical Process Control

21.1 Bottles

The champagne bottles produced by a company dedicated to the manufacture of bottles and glass containers are the products of most impact in the turnover of that company. These bottles have specifications for: height, core diameter, external diameter of the mouth, inside diameter of the mouth, weight, horizontality, verticality, ovalization and thickness of the glass.

The presence of some quality problems highlights the need to launch a project for improvement. A preliminary study shows that 85% of the times that the bottle is out of the tolerance range, it is because of weight (too less weight is equivalent to lack of resistance) or ovalization. Thus, it is decided to focus actions for improvement on both of these aspects.

The process starts with mixing raw materials (silicon, sodium carbonate, and calcium; sometimes also glass for recycling) and continues with its merger, dripping, annealing, cold treatment, and palletizing. It is known, however, that the type of defects to attack can only occur during the dripping phase.

Industrial Statistics with Minitab, First Edition. Pere Grima Cintas, Lluís Marco-Almagro and Xavier Tort-Martorell Llabrés.
© 2012 John Wiley & Sons, Ltd. Published 2012 by John Wiley & Sons, Ltd.

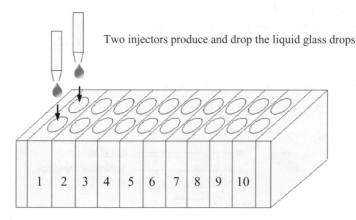

Two injectors produce and drop the liquid glass drops

The dripping phase consists of introducing a drop of liquid glass (which will then be the bottle) into the cavity of the mould. The cavities are grouped in boxes of 20, which in turn consist of 10 sections with two cavities each. Each of these two cavities is fed by an injector, according to the scheme displayed on the previous page.

It is known that the weight and/or ovalization measures lie outside of the tolerance range due to the presence of excessive variability. However, the origin of such variation is unknown. It may be due to positional aspects (an injector may be wrongly adjusted, a mould may have moved, the cooling system – which is independent for each section – may not work correctly in a certain section). It can also be due to intrinsic variability (short-term capability) or that some of the system parameters move off centre over time.

To identify the origin of the problem, a certain working day a plan for an intensive data collection is carried out. Each hour, all bottles produced in 5 consecutive boxes are collected (5 boxes × 20 bottles/box = 100 bottles). As there are seven different collection times (the first after the first working hour), we do have data on a total of 700 bottles. The collected data are in the file named BOTTLE.MTW, which is organized as follows:

Column	Variable	Label
C1	Sample	Sample number. There is a total of 7
C2	Box	Number of the box. There are 5 in each sample
C3	Section	Section. There 10 in each box
C4	Drop	Drop. A or B
C5	Weight	Bottle weight, in grams
C6	Ovality	Ovalization. Difference between max. and min. diameter, in mm

Which conclusions can be drawn from the analyses of these data?

Multi-vari charts are the most adequate tool to identify the sources of variability in a case as the one we are dealing with. First, proceed to study the weight (response variable) with all other available variables: **Stat > Quality Tools > Multi-Vari Chart**

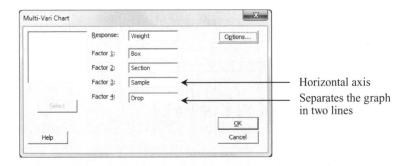

Horizontal axis
Separates the graph
in two lines

Multi-Vari Chart for Weight by Box - Drop

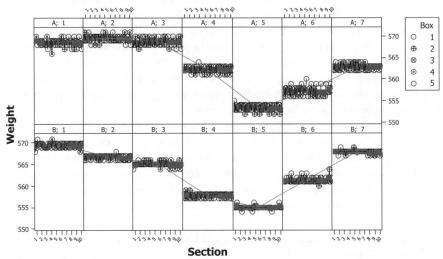

Panel variables: Drop, Sample

The most important source of variability for the weight is the one produced by the passage of time. We can hardly perceive any differences neither among sections nor among the 5 consecutive boxes of each sample. However, there seems to be a certain difference between drops. To see that more clearly, we draw the following graph without the representation of the 5 consecutive boxes.

Stat > Quality Tools > Multi-Vari Charts

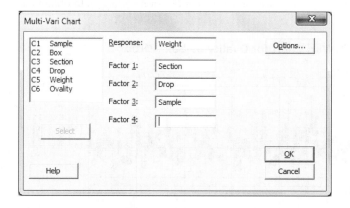

In the graph obtained, we can clearly observe the difference among samples (there is a difference of about 15 grams between the mean values of the first and fifth sample) and between drops, even though of less magnitude. In sample 7, there is a difference of about 5 grams between the mean values of A and B.

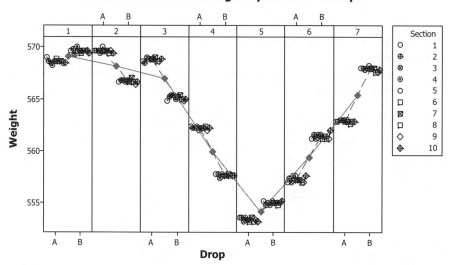

Multi-Vari Chart for Weight by Section - Sample

Panel variables: Sample

Focus now on the study of the ovality variable. Enter all factors, in the same order previously used for variable weight (box, section, sample, drop), to obtain the following chart:

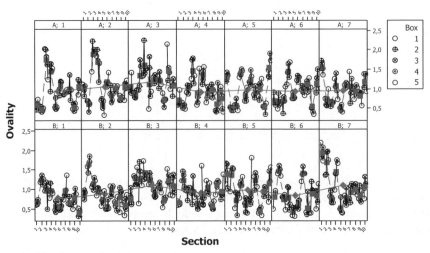

Multi-Vari Chart for Ovality by Box - Drop

Panel variables: Drop, Sample

In this case, the most important source of variability is not due to the passage of time, but it seems to be due to either the difference among sections or the difference between drops.

Remove the sample as source of variation and place section as the last factor so that it appears on the horizontal axis:

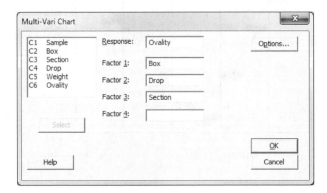

Multi-Vari Chart for Ovality by Box - Section

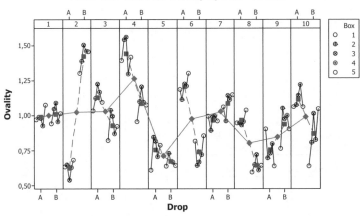

Panel variables: Section

Although the most relevant information has already been shown, other graphs can be made, too, changing the place of the factors in order to detect other aspects of the data behaviour.

As can be seen, multi-vari charts provide a very useful insight into the variability components and thus, to determine variability causes and measures to fight it.

21.2   Mattresses (1st Part)

A manufacturing process of springs for mattresses produces 3000 of such type of springs per hour, working continuously from 7 a.m to 10 p.m. As part of a Six Sigma improvement project, the decision is taken to implement a statistical control of this process. For this, every three hours a sample of 4 springs is taken and their length is measured by applying a force of 1 kg. According to the specifications, under these conditions, the length should be between 13.5 and 14.5 cm.

The data corresponding to the first week are in columns named Length 1, Time 1, and Day 1 of file MATTRESSES.MTW (the other three columns will be used in the second part of this case).

We are interested in which conclusions can be drawn from the $\bar{X} - R$ control chart constructed using the data of the first week, and whether we deal with a capable process.

To draw the $\bar{X} - R$ chart, do the following.

Stat > Control Charts > Variables Charts for Subgroups > Xbar–R

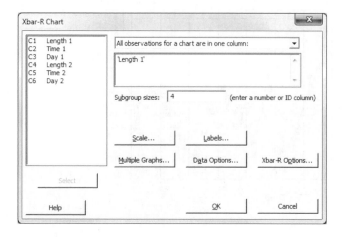

In the graphs obtained, two special causes are clearly identified. One is periodic: every day at 7 a.m., the springs are too long (they are only slightly compressed). The other cause is linked to an increase of the variability and of the average on the last day. In addition, something unusual happens the morning of the second day (a punctual increase in variability).

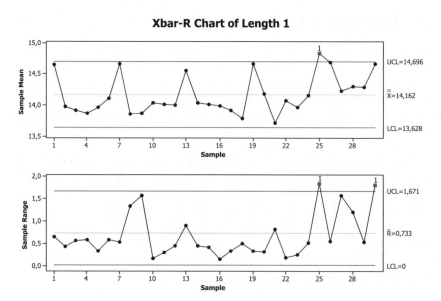

Divide the graph by days to visualize more clearly that the problem occurs at the first hour of the morning.

From button **Xbar-R Options,** in tab **Stages: Define stages. . .:** 'Day 1'

Xbar-R Chart of Length 1 by Day 1

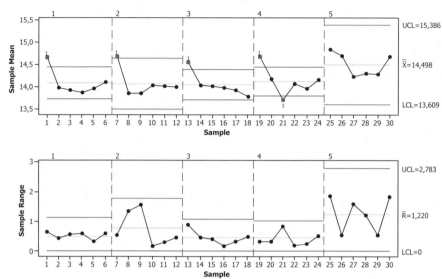

The horizontal axis has been changed to identify more easily to which sample each point corresponds (double-click on any value of the abscissa and in the dialog box that appears mark **Specify tick positions**: 1:30/1).

It is possible to recalculate the limits excluding the points corresponding to 7 a.m. of each day (there are problems at the start) and all values corresponding to Friday. From button **Xbar-R Options,** in tab **Estimate:**

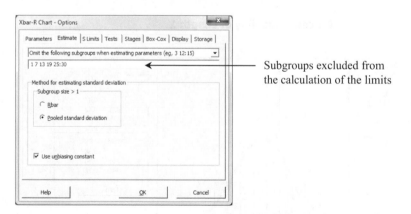

Subgroups excluded from the calculation of the limits

 Variable 'Day 1' must be removed from **Stages**, because Minitab cannot calculate the control limits for Friday since all values of this weekday were removed.

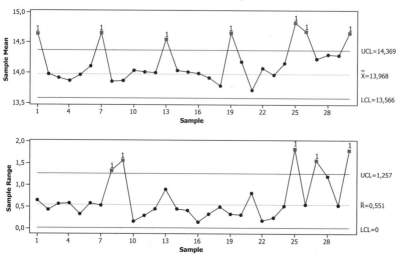

The control limits are quite narrower, and the graphs reveal more clearly the patterns described above.

Concerning the capability of the process, it does not make sense to carry out the capability study when the process is out of control. Nonetheless, we can take a look at the current situation by doing the following.

Stat > Quality Tools > Capability Analysis > Normal

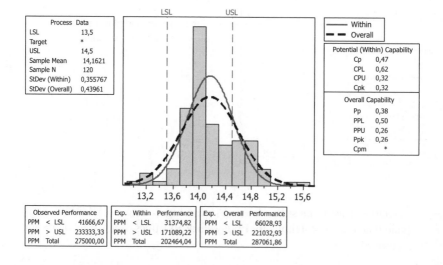

What is evident when comparing the histogram with the specifications is that a large number of springs are produced outside the tolerance limits.

21.3   Mattresses (2nd Part)

After removing the special causes – mainly due to problems with the temperature in the annealing furnace (at the first hour of the morning, the necessary value was not yet reached) and to a change of supplier – the data collection is continued the following week and the values of the variables named Length 2, Time 2, and Day 2 are obtained.

We want to know whether the problems have been solved, whether the process is in a state of control and whether it is capable.

With these data and all default options, draw again a $\bar{X} - R$ chart:

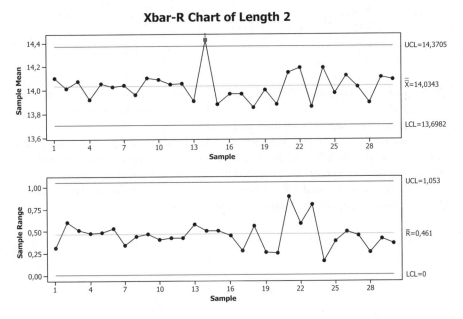

Xbar-R Chart of Length 2

The process is, basically, in a state of control. Nonetheless, the point detected outside the control limits needs to be investigated in order to find the cause that produced that behaviour. For a more detailed study, use the **Capability Sixpack** tool.

Stat > Quality Tools > Capability Sixpack > Normal

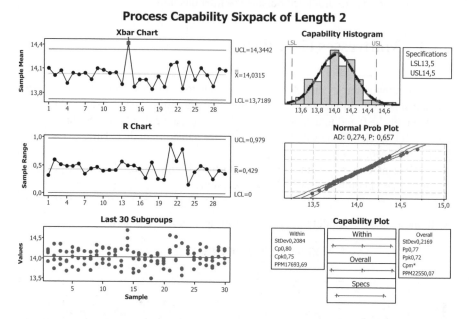

Clicking on **Options**, enter 30 in **Number of subgroups display: Last** so that the points of all samples are displayed in the graph named **Last 30 subgroups**, in such a way that they are aligned with their means and ranges.

The data follow a normal distribution reasonably well and the process is, basically, in state of control. Both the indexes and the **Capability Histogram** indicate that it is not capable and that springs are produced that do not meet the specifications.

To obtain more detailed information, carry out the following analysis: **Stat > Quality Tools > Capability Analysis > Normal**

Process Capability of Length 2

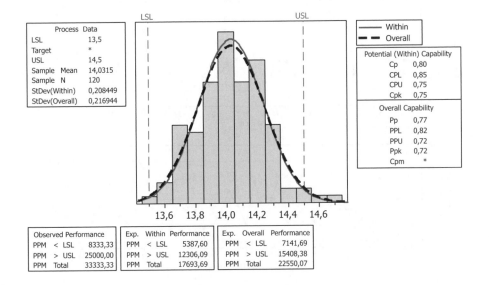

21.4 Plastic (1st Part)

Plastic parts are produced by injection and are supplied to a manufacturer of consumer electronics devices. These parts are used as covers for audio and video equipment, and among the references there is a length whose specifications are 452 mm ± 0.5 mm. This is of extreme interest to the client and has been giving problems.

As part of a project to solve this problem, it is decided to control this length by means of a $\bar{X} - R$ chart. Each hour, a sample is taken consisting of 4 consecutive parts (the working schedule is from 7 a.m. to 3 p.m., the first sample is taken at 8 a.m.). The data are in file PLASTIC_1.MTW.

Which conclusions can be drawn from the $\bar{X} - R$ chart? For this, using all default options, choose:

Stat > Control Charts > Variables Charts for Subgroups > Xbar–R. All observations for a chart are in one column: Length. **Subgroup sizes: 4.**

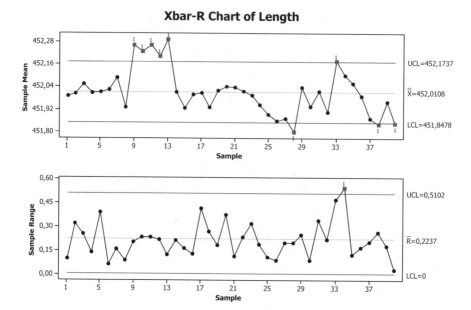

The process is clearly out of control. The R chart shows only one point outside the control limits (the rest does not seem to indicate signs of special causes).

The graph of the sample shows a nonrandom behaviour. During the first hours of the second day (around sample 10), there is a clear increase in the length (we should try to find out what is causing that). From sample 20 through 28, there is a clear decrease in the average followed by a sharp increase and a new decline. It is a complicated pattern that seems to indicate that there is a complex system of special causes affecting the length.

21.5   Plastic (2nd Part)

In addition to the values of the length being controlled, every hour, together with the extraction of the subgroup, measures of the pressure, viscosity, and temperature of injection are also taken. This information is contained in file PLASTIC_2.MTW.

The questions raised by the improvement team are the following: Do these data allow the identification of any special cause that may affect the variable length? Under which conditions of pressure, viscosity and temperature does it seem to be reasonable to inject? Is it possible to make the process capable?

The content of this file (PLASTIC_2.MTW) has the following appearance:

	C1-D	C2-D	C3	C4	C5	C6	C7	C8	C9	C10	C11	C12	C13
	Date	Time	Pressure	Viscosity	Temperature								
1	04/11/2002	8:00	830.8	41.8	118.4								
2	04/11/2002	9:00	814.1	41.8	120.7								
3	04/11/2002	10:00	818.3	43.0	120.2								
4	04/11/2002	11:00	820.4	42.1	120.3								
5	04/11/2002	12:00	808.3	42.7	120.6								
6	04/11/2002	13:00	827.0	42.8	121.9								
7	04/11/2002	14:00	821.9	41.3	123.1								
8	04/11/2002	15:00	823.0	42.2	119.8								
9	05/11/2002	8:00	819.9	42.1	123.0								

It is helpful for the analysis to have the mean and range of each subgroup in the same worksheet. To obtain this, open worksheet **PLASTIC_1.MTW** and do the following.

Calc > Make Patterned Data > Simple Set of Numbers

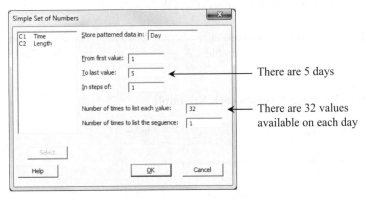

There are 5 days

There are 32 values
available on each day

Stat > Basic Statistics > Store Descriptive Statistics

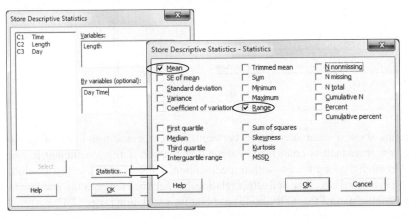

Hence, worksheet PLASTIC_1.MTW now looks as follows:

	C1-D	C2	C3	C4	C5-D	C6	C7	C8	C9	C10	C11	C12	C13
	Time	Length	Day	ByVar1	ByVar2	Mean1	Range1						
1	8:00	452.04	1	1	8:00	451.990	0.097879						
2	8:00	451.97	1	1	9:00	452.000	0.319152						
3	8:00	451.94	1	1	10:00	452.056	0.254769						
4	8:00	452.02	1	1	11:00	452.006	0.136741						
5	9:00	451.91	1	1	12:00	452.010	0.388686						
6	9:00	451.98	1	1	13:00	452.025	0.063160						
7	9:00	452.22	1	1	14:00	452.088	0.158258						
8	9:00	451.90	1	1	15:00	451.928	0.087815						
9	10:00	452.22	1	2	8:00	452.266	0.205453						

The last thing to do is to copy columns C6 and C7 and add them to the worksheet named Plastic_2 (just by using copy and paste).

To highlight all the values of a column (for example, in order to copy them) simply click on the column number (C1, C2, ...)

By means of bivariate diagrams we can study whether there is any relation between the means or ranges and the process variables. A good way to do so with a single graphical window is as follows.

Graph > Matrix Plot: Each Y versus each X: Simple

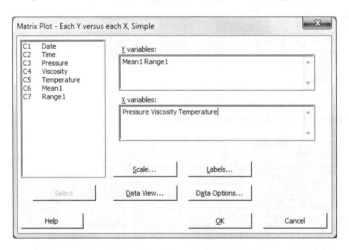

The graphs show a clear correlation between the temperature and the subgroup average length (which is confirmed analytically: the correlation coefficient is $r = 0.79$). Although the length stays within specification, it shows quite a large variation; assuming that there is a cause and effect relationship between temperature and length we can attribute part of this to the wide temperature's variation range ($7°C$).

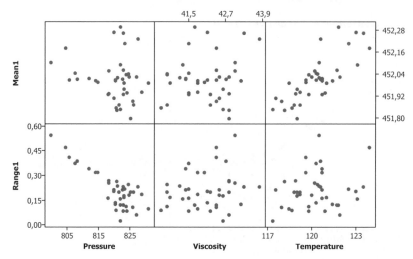

The diagram of Range versus Pressure shows that the ranges increase (which is equivalent to an unstable length) when pressure is low. On the other hand, when pressure is high, there does not seem to be an influence of pressure on the length variability. The remaining variables do not seem to affect the stability of the length.

To analyze these diagrams with more detail, do the following.

Graph > Scatterplot: Simple

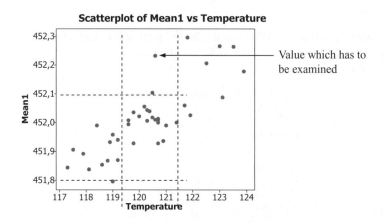

Note that if the temperature is kept between 119°C and 121°C, a condition which seems easy and cheap to meet with today's technical tools, a practically stable process (without any special causes, except for that value which needs to be studied) is obtained and with little variation (much less than the one required by the tolerances).

Analyze also the graph Range vs. Pressure:

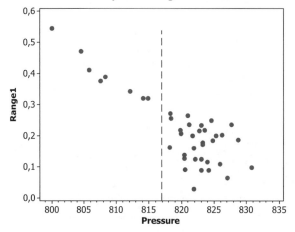

Scatterplot of Range1 vs Pressure

It seems clear that the pressure must be maintained above 817 kg/cm^2 since then no high variation in the length is caused.

Lines between numbers on the abscissa have been added by double-clicking on that axis: **Scale: Minor ticks, Number** 4; **Show** (new tab) mark **Minor ticks** in **Low.**

To carry out the capability study only with the two subgroups that meet the conditions imposed requires certain data management operations. One way of doing this is the following:

1. In worksheet PLASTIC_2.MTW, use the logic function (via **Calculator**) to identify which subgroups meet the conditions imposed.
   Calc > Calculator

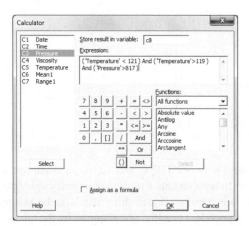

2. To select the values of the subgroups of interest, copy the values of C8 (select and copy).

3. Go to worksheet PLASTIC_1.MTW and use: **Calc > Make Patterned Data > Arbitrary Set of Numbers** in order to create a new column (coincidentally, it will also be C8) in which each value we have copied appears 4 times in a row:

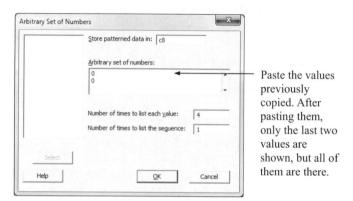

Paste the values previously copied. After pasting them, only the last two values are shown, but all of them are there.

4. Copy column Length (C2) into an empty column of the worksheet, but click first on **Subset the data**... to specify condition C8=1.

   Now you can carry out the capability study with the values of the new column which only correspond to the subgroups that meet the conditions imposed. Use the tool **Capability Sixpack** to obtain:

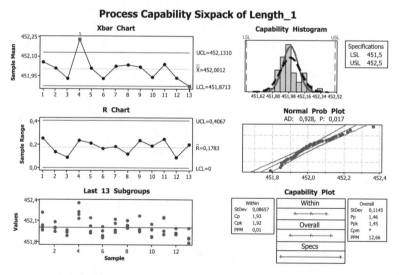

There are two points in the **Xbar Chart** (representing the sample means) beyond the control limits. Nonetheless, we can affirm that under the conditions imposed, the process is perfectly capable.

Part Five

REGRESSION AND MULTIVARIATE ANALYSIS

One of the most common tasks in data analysis is to study – and characterize if it is the case – the possible relationship between two variables, X and Y. Naturally the first step in this endeavour is to plot one variable versus the other – scatterplot – and judge if there is a pattern and what it may mean.

If there is a linear relationship between them, a not unusual case, the strength of this relation can be measured by the correlation coefficient. It is a measure easy to interpret; its value varies between -1 (perfect negative correlation, increasing X decreases Y) and 1 (perfect positive correlation, increasing X increases Y). Some remarks on their use and interpretation:

- The correlation coefficient only measures the degree of linear relationship. Two variables can be perfectly related, but if the relation is, for example, quadratic, the correlation coefficient could be very low.

- Correlation does not imply cause and effect relationship. Two variables may be highly correlated (have a high correlation coefficient), but not directly dependent on each other. There are clear and funny examples to highlight these situations: the number of firefighters who come to fight a fire and the damage caused by it (usually the more firefighters the bigger damage, but firefighters do not cause the damage; there is a third hidden variable related to these two, in this case the magnitude of the fire). This may happen in other cases, and is quite common in industrial settings.

- A correlation coefficient different from zero does not imply that the correlation exists. To get a correlation coefficient exactly equal to zero is a question of chance, a perfect balance of the data that in practice never happens. If we get Minitab to generate two columns of random numbers and to calculate the correlation coefficient between them, it will certainly not be zero. That is the reason to check if the value obtained is 'statistically significant', meaning that it is highly unlikely to obtain a value like this or bigger by chance if the samples were independent. One measure of this probability is given by the p-value, the smaller the less likely.

Once the relationship between variables has been verified it may be of interest to model it. The aim is to find the equation that best explains the relationship between X (usually called the independent variable) and Y (usually called the dependent variable). We call this a regression model or a regression equation. It can be simple, if it contains a single explanatory variable, or multiple, if it contains more than one.

The idea behind simple regression is conceptually easy: it comes to find the straight line that best fits the dots in the scatterplot of Y versus X; in this case 'best' means the one that minimizes the sum of squares of the distance between the points and itself. By the way, these distances are called the residuals, so the idea is to minimize the residuals' sum of squares. Minitab does the calculations for us – fits the model – and in addition it provides several measures of goodness of fit. They should be taken into account because, naturally, Minitab always provides an equation and it looks the same independently of whether there is a very strong or very weak relationship. The most popular measure of adjustment quality is called Determination Coefficient (R^2) and is equal to the square of the correlation coefficient between X and Y; therefore, their values are between 0 and 1, or 0 to 100 if, as Minitab does, it is expressed as a percentage. However, the most important model (goodness of fit) checking activity is to look at the residuals; they must behave as an independent random sample from a normal distribution centred in zero. The checking is conducted through several residual plots.

When there are several independent variables – variables that may explain the behaviour of the response – finding the 'best' model is not an obvious task. The main complication is to decide which X variables should be included in the model, and in which metric. Minitab provides the possibility to use different strategies to select the variables to be included in the model. The simplest one is to fit all possible models and order them by a set of goodness of fit criteria (best subset). With this information it is easy to choose 2 or 3 models to be studied in more detail (influential points, outliers, residual checking. . .). In the multiple liner regression context, model checking is especially relevant to help build a good model and to understand the relationship between the dependent and independent variables.

Before finishing the issue of multiple regression, a couple of warnings:

− Whenever the number of independent (X's) variables is similar to the number of data points available, the model fitted will explain the response very well; this is a delusion, remember that a straight line is defined by two points (a model with two parameters), so there is a perfect fit independently of the type of relation between X and Y. Three points determine a plane (model with 3 parameters, etc.). That is the reason behind the Adjusted Determination Coefficient (R^2 – Adj) similar to R^2 but taking into account the relationship between the number of parameters and the number of data points.

- In general, regression is not an adequate technique to model time series data (data gathered at regular time intervals). These data will not fulfil the independence assumption; they may have trends, seasonal changes or other sorts of time dependence. When this is disregarded and a regression model is fitted, an apparently very good model may result, but model checking – a plot of residuals vs. time – will clearly show the assumptions violation. There are special tools to model time series data; Minitab provides them on the Time Series menu.

Part 5 also includes a chapter on multivariate analysis. With volumes of data available increasing at an astonishing rate, the relevance of these techniques is increasing at a similar pace. They allow us to find patterns, to group items (or individuals) with similar characteristics or to condense the information of several variables into one or two new ones. These techniques are one of the pillars of the so-called 'data mining', a semi-automatic way to discover patterns, structures and classifications in large databases.

22

Correlation and Simple Regression

22.1 Correlation Coefficient

 Consider the measures from the height (in inches) and weight (in pounds) of the 92 students whose data are included in the file PULSE.MTW.

Before starting to calculate the correlation coefficient, it is highly recommended to construct a scatterplot of the data to identify possible outliers, nonlinear relationships (correlation coefficient only measures linear relationships), etc.

Graph > Scatterplot: Simple

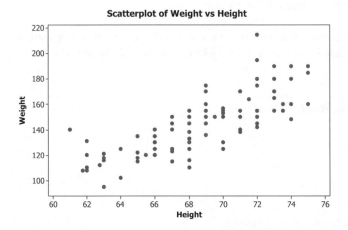

Industrial Statistics with Minitab, First Edition. Pere Grima Cintas, Lluís Marco-Almagro and Xavier Tort-Martorell Llabrés.
© 2012 John Wiley & Sons, Ltd. Published 2012 by John Wiley & Sons, Ltd.

To calculate the correlation coefficient, do the following.

Stat > Basic Statistics > Correlation

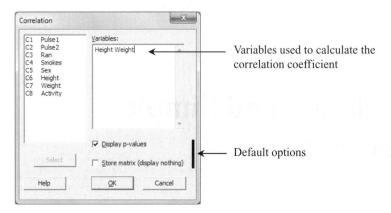

Variables used to calculate the correlation coefficient

Default options

In this way, the output in the session window looks as follows:

Correlations: Height; Weight

Pearson correlation of Height and Weight = 0.785 ◄── Correlation coeff. value
P-Value = 0.000 ◄─────────────────────── The value obtained is clearly significant

---

The statistical significance of a correlation coefficient depends not only on its value, but also on the size of the samples used for its calculation. Do not compare directly correlation coefficients obtained from data sets of different size, but through the p-value.

You can also select more than two variables, in which case pairwise correlation coefficients are calculated. For example, also include the variable named 'Pulse1' to obtain:

Correlations: Height; Weight; Pulse1

```
           Height   Weight
Weight     0.785
           0.000

Pulse1    -0.212   -0.202
           0.043    0.053

Cell Contents: Pearson correlation
                P-Value
```

Observe that just below the value of the correlation coefficient lies the corresponding p-value. For example, the correlation coefficient between 'Height' and 'Pulse1' is -0.212 and its p-value is 0.043.

 A significant correlation coefficient does not necessarily imply a cause-effect relationship.

If you check the option **Store matrix (display nothing)**, the correlations matrix is stored in a matrix with name CORRE1 and only the following appears in the session window (the option: **Editor** > **Enable Commands** must have been activated):

```
MTB > Name m1 ''CORR1''
MTB > Correlation 'Height' 'Weight' 'Pulse1' 'CORR1'.
```

 You can also store data in matrix format. M1, M2, ... are the names used by default (just like C1, C2, ... are used for columns and K1, K2, ... for constants).

To visualize the results you can write the matrix in the worksheet by doing the following.

Data > Copy > Matrix to Columns

Dialog window to convert matrices into columns

In this case, the matrix of correlations CORR1 is copied in the columns C9, C10 and C11

↓	C1	C2	C3	C4	C5	C6	C7	C8	C9	C10	C11
	Pulse1	Pulse2	Ran	Smokes	Sex	Height	Weight	Activity	CORR1_1	CORR1_2	CORR1_3
1	64	88	1	2	1	66,00	140	2	1,00000	0,78487	-0,21179
2	58	70	1	2	1	72,00	145	2	0,78487	1,00000	-0,20222
3	62	76	1	1	1	73,50	160	3	-0,21179	-0,20222	1,00000
4	66	78	1	1	1	73,00	190	1			
5	64	80	1	2	1	69,00	155	2			
6	74	84	1	2	1	73,00	165	1			
7	84	84	1	2	1	72,00	150	3			
8	68	72	1	2	1	74,00	190	2			
9	62	75	1	2	1	72,00	195	2			
10	76	118	1	2	1	71,00	138	2			

The matrix of correlations copied in the worksheet does not contain information on the p-values. The diagonal values are always equal to one, since they indicate the correlation of a variable with itself.

22.2   Simple Regression

Minitab offers many options for modelling regression equations, some of which are incorporated in version 16. For simple regression equations (there is a single explanatory variable), the most used options are:

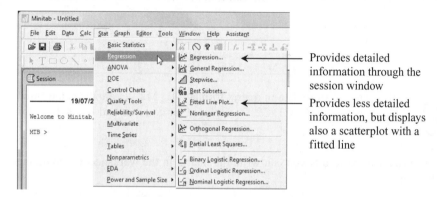

Provides detailed information through the session window

Provides less detailed information, but displays also a scatterplot with a fitted line

 Consider again the data in file PULSE.MTW (containing the height and weight of 92 students) and transform them using the following conversion rules: one inch = 2.54 cm and 1 pound = 0.454 kg.

Specifically, to convert inches to centimetres, do: **Calc > Calculator:**

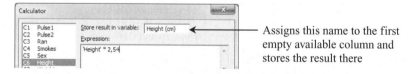

Assigns this name to the first empty available column and stores the result there

The data of height (in cms) appears with three decimal places, but they can be rounded using function **Round** within **Calculator**:

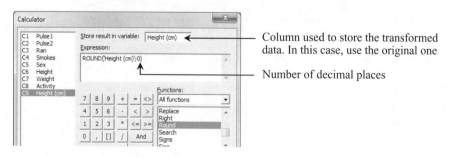

Column used to store the transformed data. In this case, use the original one

Number of decimal places

The function **Round** can also be used within an arithmetic expression. For instance, to get the weight in kilograms and with only one decimal place, do:

One decimal place

22.3   Simple Regression with 'Fitted Line Plot'

Stat > Regression > Fitted Line Plot

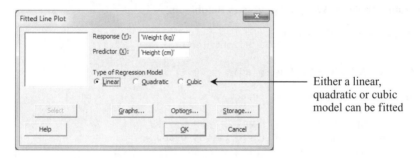

Either a linear, quadratic or cubic model can be fitted

Regression equation $\longrightarrow$

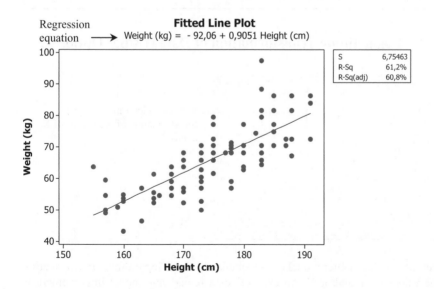

Fitted Line Plot
Weight (kg) = - 92,06 + 0,9051 Height (cm)

S	6,75463
R-Sq	61,2%
R-Sq(adj)	60,8%

The goodness of fit indicators that appear in the upper-right box of the scatterplot are:

S: Standard deviation of residuals (residual = true value of the response variable − fitted value obtained using the equation)

R-Sq Coefficient of determination (R^2): Measure of the goodness of fit. It is computed as the square of the correlation coefficient ($\times 100$, to be given in %)

R-Sq (adj): Adjusted coefficient of determination. Measure of the goodness of fit used in multiple regression (does not apply in simple regression).

The session window only adds the analysis of variance table, which can be used as a significance test for the coefficient of the explanatory variable. In our example, the coefficient of the height variable is clearly significant since the p-value = 0.000.

```
Regression Analysis: Weight (kg) versus Height (cm)

The regression equation is
Weight (kg) = -92.06 + 0.9051 Height (cm)


S = 6.75463 R-Sq = 61.2% R-Sq(adj) = 60.8%


Analysis of Variance

Source        DF       SS       MS       F       P
Regression     1   6473.4  6473.37  141.88  0.000
Error         90   4106.3    45.63
Total         91  10579.6
```

22.3.1   Logarithmic Transformation of Data (Using 'Options')

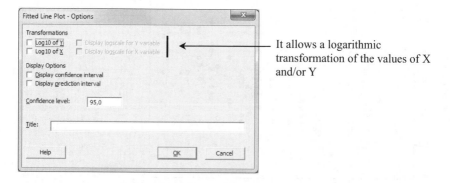

It allows a logarithmic transformation of the values of X and/or Y

Sometimes, a much better model can be obtained using a logarithmic transformation of the values of X and/or Y. An example of a remarkable model improvement is obtained by transforming both variables when creating a model that aims to explain the brain weight of mammals as a function of the total body weight.

Example 22.1: The file BRAINS.MTW contains the species names, brain weight (in g) and total body weight (in kg) of 62 species of mammals. The data are taken from S. Weisberg's book *Applied Liner Regression* (Reproduced by permission of John Wiley & Sons).

The scatterplot obtained using the original data appears below; observe that most of the data points (in total there are 62) are grouped very close to zero.

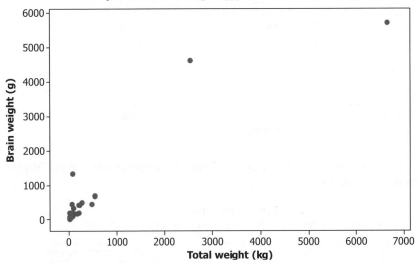

It is not correct to fit a linear equation in these circumstances since the two points that appear isolated (corresponding to the Asian and African elephant) do have a disproportionate influence on the line. One could think that a possible solution is to consider these two points as outliers which can be eliminated, but in doing so the problem is not really solved since other apparently outliers appear and the task of eliminating points will continue without finding an adequate solution (besides, if points are eliminated the model loses its generalization).

A better solution consists in transforming the original data; in this case a logarithmic transformation of the two variables provides a far better distribution of the points and a very suitable model.

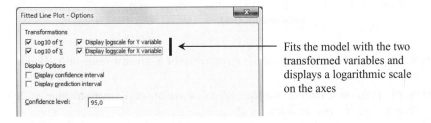

Fits the model with the two transformed variables and displays a logarithmic scale on the axes

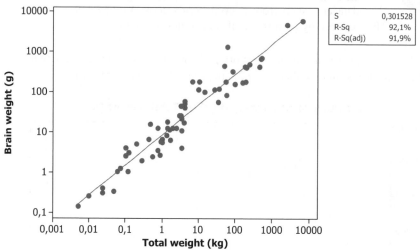

In the same window choose the options **Display confidence interval** and **Display prediction interval** to obtain:

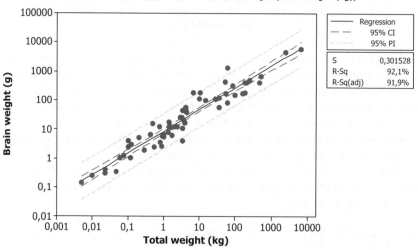

95% CI: Confidence interval for the *mean value* of Y given a value of X (Bands closer to the line).

95% PI: Prediction interval for the *individual values* of Y given a value of X (Bands farther from the line).

22.3.2   Graphical Analysis of Residuals ('Graphs')

To make sure that a good model has been fitted, not able to be improved with available data, the residuals should be analyzed to verify that they behave randomly and contain no information likely to be incorporated into the model.

 Example 22.2: The file RESIDUALS.MTW contains 50 values of X and another 50 values of Y to highlight the importance of analyzing the residuals after fitting a model.

Fit a straight line to these data to obtain:

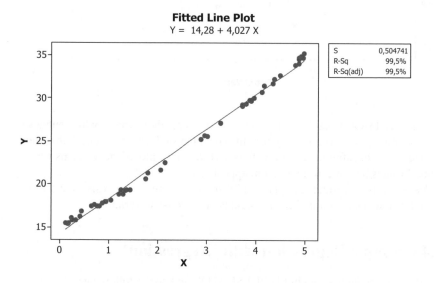

Fitted Line Plot
Y = 14,28 + 4,027 X

S	0,504741
R-Sq	99,5%
R-Sq(adj)	99,5%

Further, by creating a graph of the residuals versus the fitted values, it becomes evident that the fitted model can be improved.

Stat > Regression > Fitted Line Plot > Graphs

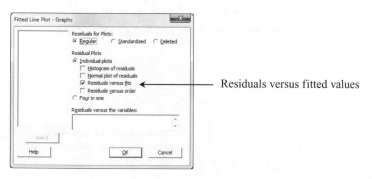

Residuals versus fitted values

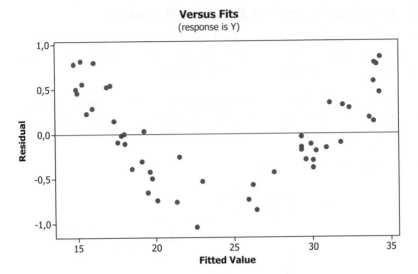

Versus Fits
(response is Y)

The residuals do not appear in a random way, but in a parabolic shape, which suggests that a quadratic model would be a better fit. The interested reader can check that such is the case, and that after the quadratic fit is performed the plot of the residuals versus the fitted values has an entirely random appearance.

The analysis of residuals is also useful to confirm the hypothesis of the fitted model, like the one for normality of residuals or its constant variance.

22.4   Simple Regression with 'Regression'

Consider once again the data in file PULSE.MTW and do the following.

Stat > Regression > Regression

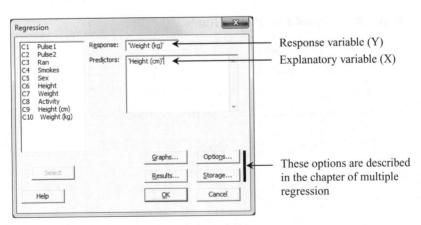

Response variable (Y)
Explanatory variable (X)

These options are described in the chapter of multiple regression

Use all default options to obtain the following (blank spaces have been added to improve the appearance of the output and facilitate its interpretation).

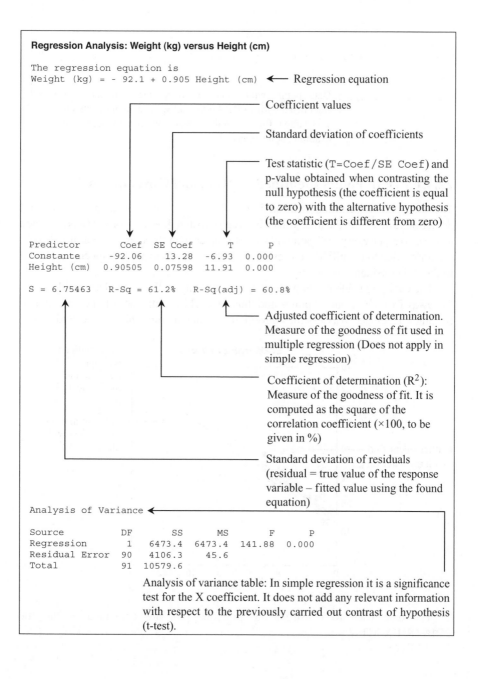

Regression Analysis: Weight (kg) versus Height (cm)

```
The regression equation is
Weight (kg) = - 92.1 + 0.905 Height (cm)
```
◄── Regression equation

Coefficient values

Standard deviation of coefficients

Test statistic ($T=Coef/SE\ Coef$) and p-value obtained when contrasting the null hypothesis (the coefficient is equal to zero) with the alternative hypothesis (the coefficient is different from zero)

```
Predictor        Coef   SE Coef      T       P
Constante      -92.06     13.28   -6.93   0.000
Height (cm)   0.90505   0.07598   11.91   0.000

S = 6.75463    R-Sq = 61.2%    R-Sq(adj) = 60.8%
```

Adjusted coefficient of determination. Measure of the goodness of fit used in multiple regression (Does not apply in simple regression)

Coefficient of determination (R^2): Measure of the goodness of fit. It is computed as the square of the correlation coefficient ($\times 100$, to be given in %)

Standard deviation of residuals (residual = true value of the response variable – fitted value using the found equation)

```
Analysis of Variance
```
◄────

```
Source            DF        SS      MS       F       P
Regression         1    6473.4  6473.4  141.88   0.000
Residual Error    90    4106.3    45.6
Total             91   10579.6
```

Analysis of variance table: In simple regression it is a significance test for the X coefficient. It does not add any relevant information with respect to the previously carried out contrast of hypothesis (t-test).

```
Unusual Observations    ←───────────────────────────────────

  Height  Weight
Obs   (cm)   (kg)      Fit  SE Fit  Residual  St Resid
  9    183  88.500  73.560   0.953    14.940     2.23R
 25    155  63.600  48.219   1.643    15.381     2.35R
 40    183  97.600  73.560   0.953    24.040     3.60R
 84    173  49.900  64.510   0.714   -14.610    -2.18R

R denotes an observation with a large standardized residual.
```

Data points with standardized residuals greater than two (identified with an R). Additionally, the data points that have a very large influence on the line are identified with an X; in this example, none exists.

22.4.1   Data Points Identified as 'Unusual Observations'

The data points identified with an R are those that fall more than two standard deviations away from the line. Values between two and three are normal (it is expected to have five per every 100 points). Values greater than three are scarce: it is highly recommended to identify the corresponding observations to apply a special treatment to them if considered necessary.

The data points identified with an X may have a small standardized residual, but fall away from the cloud of points and thus have a large influence on the line; thus, it becomes necessary to assess the convenience of maintaining them in the study.

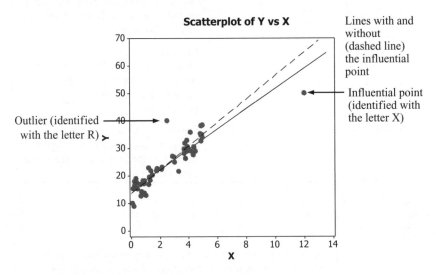

The data used to construct the previous plot are contained in the file POINTS_RX.MTW.

23

Multiple Regression

23.1   File 'CARS2'

This file, containing information of 66 cars, is organized as follows:

Column	Variable	Label
C1	Maker	Car brand
C2	Model	Model
C3	Num. Cyl.	Number of cylinders
C4	Disp.	Engine displacement (cc)
C5	Pow.	Power (CV)
C6	Max. speed	Maximum speed (Km/h)

The aim is to specify a regression model that explains the maximum speed as a function of the available information.

23.2   Exploratory Analysis

To start, the best strategy consists in taking a look at the data and a good way to do this is through: **Graph > Matrix Plot: Simple**

Industrial Statistics with Minitab, First Edition. Pere Grima Cintas, Lluís Marco-Almagro and Xavier Tort-Martorell Llabrés.
© 2012 John Wiley & Sons, Ltd. Published 2012 by John Wiley & Sons, Ltd.

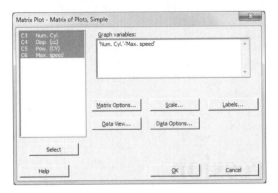

Highlight the columns of interest in the box on the left and click on **Select**

Matrix Plot of Num. Cyl.; Disp. (cc); Pow. (CV); Max. speed

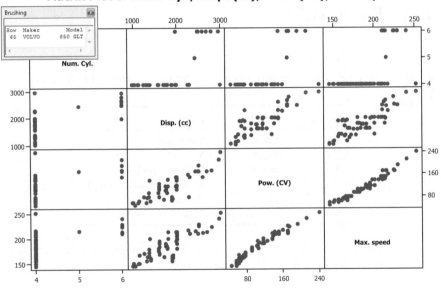

In the previous plot, observe that there is only one car with five cylinders. Use the option **Brush** (see Chapter 4) to identify that this point belongs to a Volvo 850 GLT.

Also, it seems that the relationship between power and maximum speed is not linear but quadratic. We will try to confirm this impression by analyzing the corresponding residuals.

23.3   Multiple Regression

Stat > Regression > Regression

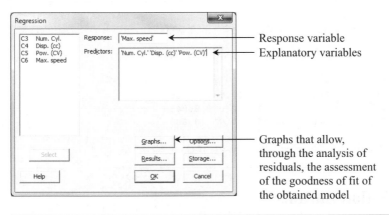

Response variable
Explanatory variables

Graphs that allow,
through the analysis of
residuals, the assessment
of the goodness of fit of
the obtained model

Regression Analysis: Max. speed versus Num. Cyl.; Disp. (cc); Pow. (CV)

```
The regression equation is
Max. speed = 135 − 2.36 Num. Cyl. − 0.00074 Disp. (cc) + 0.589 Pow.(CV)

64 cases used, 2 cases contain missing values

Predictor        Coef    SE Coef       T      P
Constant      134.518      4.496   29.92  0.000
Num. Cyl.      -2.356      1.300   -1.81  0.075
Disp. (cc)   -0.000737   0.003000   -0.25  0.807
Pow. (CV)     0.58944    0.03106   18.98  0.000
```

Values of the coefficients (`Coef`) with their standard deviation (`SE coef`),
test statistic (`T`) and p-value (p) when contrasting H_0 (the coefficient is
equal to zero) with the alternative H_1 (the coefficient is different from zero)

```
S = 5.29292 R-Sq = 95.7% R-Sq(adj) = 95.5%
```

Some measures of goodness of fit. In multiple
regression `R-Sq` loses interest and `R-Sq(adj)` is used

```
Analysis of Variance

Source          DF      SS     MS      F      P
Regression       3   37606  12535  447.45  0.000
Residual Error  60    1681     28
Total           63   39287
```

Analysis of variance table:
Joint significance test for all coefficients

```
Source      DF  Seq SS
Num. Cyl.    1   11242
Disp. (cc)   1   16275
Pow. (CV)    1   10088
```

Contribution of each explanatory variable to the sum
of squares explained by the regression

```
Unusual Observations

       Num.
Obs   Cyl.  Max. speed      Fit  SE Fit  Residual  St Resid
  5   4.00     195.000  184.041   0.778    10.959      2.09R
 33   4.00           *  167.620   2.559         *       * X
 44   4.00     252.000  264.357   2.821   -12.357     -2.76RX
 56   4.00     145.000  157.371   1.464   -12.371     -2.43R

R denotes an observation with a large standardized residual.
X denotes an observation whose X value gives it large leverage.
```

List of unusual observations. See chapter
of simple regression for details

Since the variable 'Displacement' is not significant (p-value much greater than 0.05), a new model excluding this variable is specified, as shown below:

```
The regression equation is
Max. speed = 134 - 2.46 Num. Cyl. + 0.583 Pow. (CV)

64 cases used, 2 cases contain missing values

Predictor      Coef   SE Coef       T       P
Constant    134.296     4.370   30.73   0.000
Num. Cyl.    -2.460     1.221   -2.02   0.048
Pow. (CV)   0.58341   0.01887   30.91   0.000


S = 5.25200   R-Sq = 95.7%   R-Sq(adj) = 95.6%
```

23.4   Option Buttons

23.4.1   Graphs

The following graphs allow us, through the analysis of residuals, to assess the validity of the fitted model.

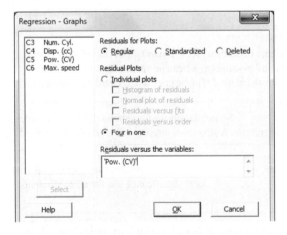

There are different types of residuals:

Regular: Raw residuals, computed as the difference between the true value and the corresponding fitted value.

Standardized: Computed as the regular residual divided by the standard deviation of the residuals.

Deleted: Similar to the **Standardized**, but in each case the data point whose residual is to be computed is excluded in the estimation of the coefficients of the model.

The requested graphs are **Four in one**, which displays in one graphical window the four more relevant residual plots and the graph of residuals versus the variable Power, since as previously illustrated, it seems that the relationship between power and maximum speed is not linear but quadratic.

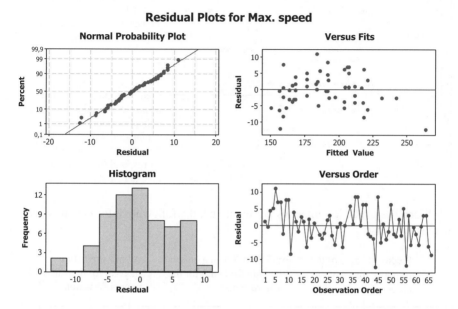

The shape of the plot of residuals versus fitted values (cloud of points in a parabolic shape) suggests the possibility of improving the model by transforming a variable. The plot of the residuals versus the variable Power suggests the convenience of a transformation of this variable: for instance taking the square of it.

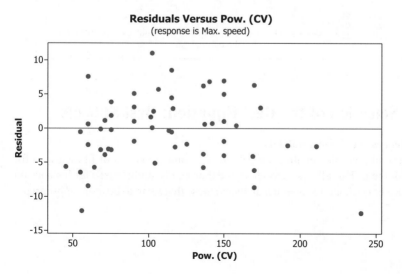

23.4.2   Options

The most common ones are:

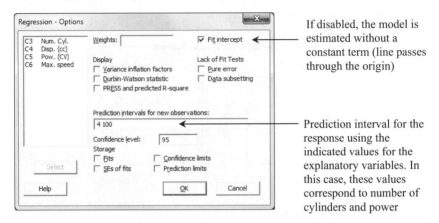

If disabled, the model is estimated without a constant term (line passes through the origin)

Prediction interval for the response using the indicated values for the explanatory variables. In this case, these values correspond to number of cylinders and power

When the prediction intervals are requested, the following information appears in the output list:

Predicted value of the response for the new observation

A 95% confidence interval (the indicated one) for the mean response

A 95% prediction interval for the single new observation

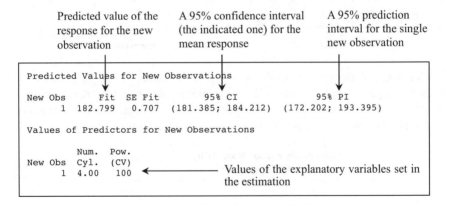

```
Predicted Values for New Observations

New Obs      Fit   SE Fit         95% CI              95% PI
       1  182.799   0.707  (181.385; 184.212)  (172.202; 193.395)

Values of Predictors for New Observations

             Num.   Pow.
New Obs      Cyl.   (CV)
       1     4.00    100
```

Values of the explanatory variables set in the estimation

23.5   Selection of the Best Equation: Best Subsets

Stat > Regression > Best Subsets
It generates all possible equations, together with some goodness of fit measures for each one of them. This allows us to select and thoroughly study (based for instance on significance of coefficients, residual analysis) those that seem to be more appropriate.

In the following, the square of the variable Power, named Pow2, is entered as a new explanatory variable.

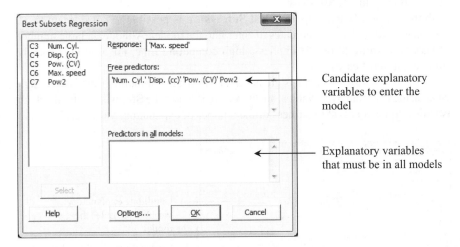

Candidate explanatory variables to enter the model

Explanatory variables that must be in all models

The list that appears, with all default options (the best two equations with one to four explanatory variables), is:

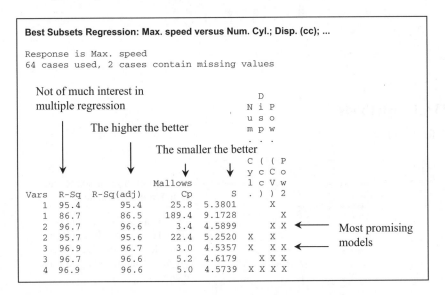

Best Subsets Regression: Max. speed versus Num. Cyl.; Disp. (cc); ...

```
Response is Max. speed
64 cases used, 2 cases contain missing values
```

Not of much interest in multiple regression

The higher the better

The smaller the better

					D
					N i P
					u s o
					m p w
					. . .
					C ((P
					y c C o
			Mallows		l c V w
Vars	R-Sq	R-Sq(adj)	Cp	S	.)) 2
1	95.4	95.4	25.8	5.3801	X
1	86.7	86.5	189.4	9.1728	X
2	96.7	96.6	3.4	4.5899	X X
2	95.7	95.6	22.4	5.2520	X X
3	96.9	96.7	3.0	4.5357	X X X
3	96.7	96.6	5.2	4.6179	X X X
4	96.9	96.6	5.0	4.5739	X X X X

Most promising models

The selection of the best equation depends also on aspects such as the simplicity of the model or its ease of interpretation. In many cases, it is questionable to say that we have the 'best' model. Thus, perhaps it would be better to talk about a 'good' model.

23.6   Selection of the Best Equation: Stepwise

Stat > Regression > Stepwise

With the previously used option (**Best Subsets**), it is possible to consider up to 31 explanatory variables as candidates to enter into the model. Although a large number of variables (from 20 or 25) implies a high computing time, this option is usually sufficient in most cases.

If the number of variables is greater than 31, or the computing time is very high when using less than 31 variables, an alternative is to use the **Stepwise** method, which is known to be very fast regardless of the number of explanatory variables.

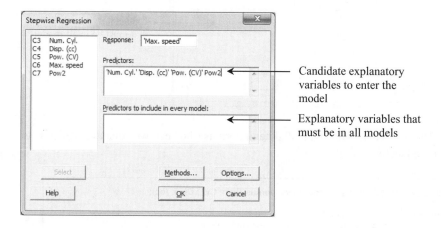

Candidate explanatory variables to enter the model

Explanatory variables that must be in all models

23.6.1   Methods

Default values:

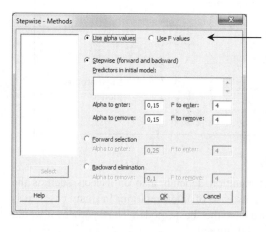

Selection of the criterion to enter (or to remove) variables

Alpha values: The variable is entered (or removed) into (from) the model if the p-value is less (greater) than the indicated value of alpha

F values: The variable is entered (or removed) into (from) the model if the square of the t-ratio is greater (or less) than the indicated F-value

Stepwise: The most common method. The variables are entered into the model or removed from the model at each step. It is possible to choose a set of variables to be in the model when the process starts (**Predictors in initial model**). By default, the initial model does not include any variable.

Forward selection: Once the variables are included in the model they cannot be removed.

Backward elimination: The variables that have been included in the model can be removed (all variables initially included in the model), but once a variable is removed, it never enters again.

Use all default options to obtain:

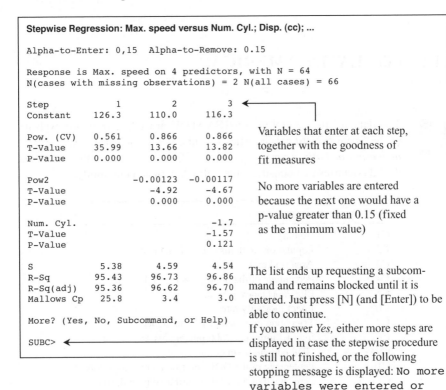

```
Stepwise Regression: Max. speed versus Num. Cyl.; Disp. (cc); ...

Alpha-to-Enter: 0,15   Alpha-to-Remove: 0.15

Response is Max. speed on 4 predictors, with N = 64
N(cases with missing observations) = 2 N(all cases) = 66

Step            1       2        3
Constant    126.3   110.0    116.3

Pow. (CV)   0.561   0.866    0.866
T-Value     35.99   13.66    13.82
P-Value     0.000   0.000    0.000

Pow2             -0.00123 -0.00117
T-Value           -4.92    -4.67
P-Value            0.000    0.000

Num. Cyl.                   -1.7
T-Value                     -1.57
P-Value                     0.121

S            5.38    4.59     4.54
R-Sq        95.43   96.73    96.86
R-Sq(adj)   95.36   96.62    96.70
Mallows Cp   25.8    3.4      3.0

More? (Yes, No, Subcommand, or Help)

SUBC>
```

Variables that enter at each step, together with the goodness of fit measures

No more variables are entered because the next one would have a p-value greater than 0.15 (fixed as the minimum value)

The list ends up requesting a subcommand and remains blocked until it is entered. Just press [N] (and [Enter]) to be able to continue.
If you answer *Yes*, either more steps are displayed in case the stepwise procedure is still not finished, or the following stopping message is displayed: No more variables were entered or removed

 You must enter the requested subcommand when obtaining the result of a stepwise regression. Minitab remains blocked (all menus are disabled) until an answer is given.

24

Multivariate Analysis

24.1 File 'LATIN_AMERICA'

The data in file LATIN_AMERICA.MTW are taken from the website of the Spanish National Statistics Institute (www.ine.es). Inebase: *Social indicators of Latin American countries 1998*. It contains information on the 22 countries that constitute the Latin American community:

Column	Content
C1	Country
C2	Population (in thousands of inhabitants)
C3	Area (in km^2)
C4	Percentage of the population under 15 years
C5	Life expectancy at birth
C6	Infant mortality rate
C7	Telephone lines per 1000 inhabitants
C8	Internet users per 1000 inhabitants
C9	GDP per capita (in dollars)
C10	Percentage of the GDP contributed by agriculture
C11	Percentage of the GDP contributed by industry
C12	Percentage of the GDP contributed by services

The aim is to classify or organize the countries according to the information available.

Industrial Statistics with Minitab, First Edition. Pere Grima Cintas, Lluís Marco-Almagro and Xavier Tort-Martorell Llabrés.
© 2012 John Wiley & Sons, Ltd. Published 2012 by John Wiley & Sons, Ltd.

24.2   Principal Components

Statistical method used to compute new variables function of the original ones. These new variables are called 'principal components' and it is expected that few of them contain most of the information in the data.

Stat > Multivariate > Principal Components

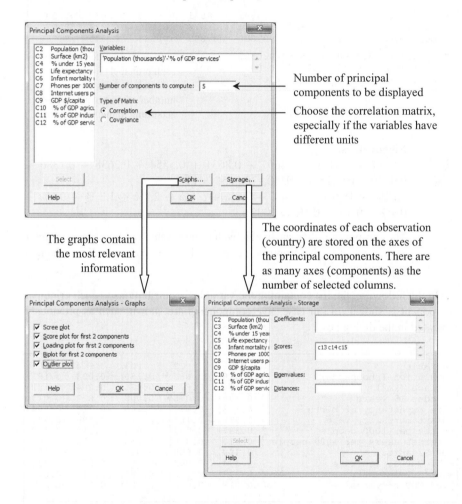

Number of principal components to be displayed

Choose the correlation matrix, especially if the variables have different units

The graphs contain the most relevant information

The coordinates of each observation (country) are stored on the axes of the principal components. There are as many axes (components) as the number of selected columns.

In case the covariance matrix is chosen, the sum of the eigenvalues is equal to the sum of the variables' variances. In case the correlation matrix is chosen, the data will be normalized and the sum of the eigenvalues will be equal to the number of variables. In either case, the eigenvalues represent each component's contribution to the explanation of the variability of the data.

The first part on the shown output list provides information on the magnitude of the eigenvalues, ordered from highest to lowest, on the proportion they represent with respect to the total (proportion of the global variability explained by this component), and on the cumulative proportion.

Principal Component Analysis

```
Eigenanalysis of the Correlation Matrix

Eigenvalue  5.5117  2.0441  1.4691  0.8631   0.5554   0.2638   0.1386
Proportion  0.501   0.186   0.134   0.078    0.050    0.024    0.013
Cumulative  0.501   0.687   0.820   0.899    0.949    0.973    0.986

Eigenvalue  0.0660  0.0475  0.0350  0.0056   Eigenvalues associated to
Proportion  0.006   0.004   0.003   0.001    each principal component
Cumulative  0.992   0.996   0.999   1.000    (in total, as many as the
```
number of variables) ordered according to their importance.

Eigenvalues:
5.5117 + 2.0441 + 1.4691 + 0.8631 + 0.5554 + 0.2638 + 0.1386 + 0.0660 + 0.0475 + 0.0350 + 0.0056 = 11
50.1% + 18.6% + 13.4% + 7.8% + 5.0% + 2.4% + 1.3 % + 0.6% + 0.4% + 0.3% + 0.1% = 100%

The proportion that each eigenvalue represents with respect to their sum is equal to the proportion explained by the corresponding principal component.

In the following, the list containing the variables' contributions to each principal component is shown. In our case, five components are shown, such as previously indicated in the dialog box.

Variable	PC1	PC2	PC3	PC4	PC5
Population (thousands)	0.016	0.667	0.150	0.023	0.191
Surface (km2)	-0.024	0.679	0.076	0.004	0.122
% under 15 years	-0.398	-0.076	0.008	0.073	0.013
Life expectancy at birth	0.358	-0.157	0.140	-0.125	0.564
Infant mortality rate	-0.370	0.162	-0.111	0.096	-0.487
Phones per 1000 people	0.387	-0.033	0.010	0.266	-0.320
Internet users per 1000 people	0.310	0.030	0.053	0.625	0.045
GDP $/capita	0.380	0.085	0.018	0.235	-0.352
% of GDP agriculture	-0.334	-0.093	-0.062	0.561	0.330
% of GDP industry	0.272	0.122	-0.555	-0.314	-0.067
% of GDP services	0.019	-0.066	0.791	-0.197	-0.228

The first component consists essentially of the contribution of the variables related to development

The second component is related to the size of the country

The third is focused on the distribution of the GDP by industry and services

The first graph is a kind of Pareto diagram of the eigenvalues that allows the visual assessment of the importance of each of the components.

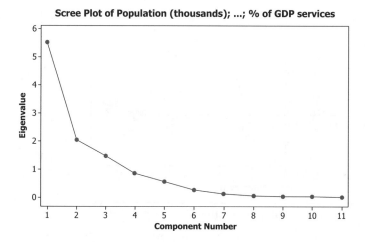

The second graph represents each of the observations (in our case, countries) on the coordinates of the first two components. To identify which country corresponds to each point, you can use the **Brush** option.

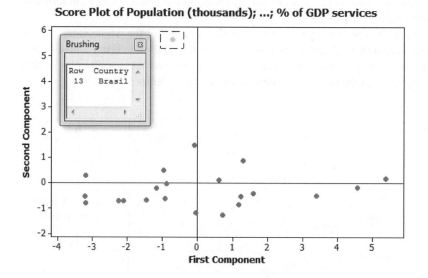

An alternative is to edit the graph adding labels to each point. To do so, right-click on the graph and choose **Add** > **Data Labels: Use labels from column**: Country.

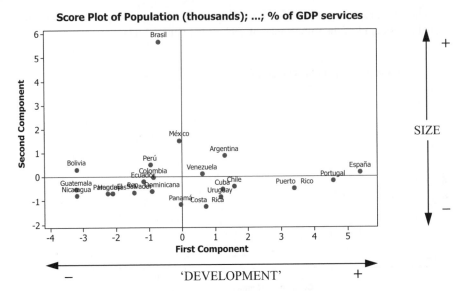

It is not always possible to give the principal components a name that reflects well the physical meaning of the group of variables that compose them. In this case, you can see in the list of each variable's contribution to the principal components that the first of these is especially affected by variables related to what might be called 'development'. The second component is related to the country's size (in terms of both area and population).

As seen, the third component, which explains 13.4% of the whole variability of the data, is basically related to the distribution of the GDP in industry and services. As the coordinates of this third component have been saved, a scatterplot of this component versus the first can be represented as shown below:

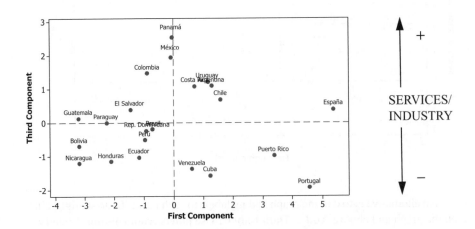

The third graph shows the variables in the coordinates that correspond to the values in the two principal components:

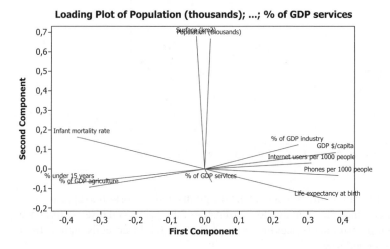

Note that Area and Population stand out because of their large value corresponding to the second component, which is related to the country's size. Concerning the first component, related to what could be termed 'development', it can be seen which variables have a positive and which a negative component.

If wanted it is possible to construct diagrams like this (though without the lines) for any other pair of components. Simply save the coefficients in the dialog box **Storage: Coefficients** and add a column with the name of the variables (in the same order of the columns) in order to assign a label to each point.

Principal Components Analysis - Storage

		Coefficients:	c16 c17 c18
C2	Population (thou		
C3	Surface (km2)		
C4	% under 15 yeaı		
C5	Life expectancy	Scores:	c13 c14 c15
C6	Infant mortality ı		
C7	Phones per 100C		
C8	Internet users pı		
C9	GDP $/capita	Eigenvalues:	
C10	% of GDP agricı		
C11	% of GDP indus		
C12	% of GDP servic	Distances:	
C13	First Componen		

	C16	C17	C18	C19-T
	Coeff First Comp.	Coeff Second Comp.	Coeff Third Comp.	Variables
	0,015642	0,667248	0,149828	Population
	-0,023823	0,679000	0,076497	Surface
	-0,397857	-0,076350	0,008033	%<15 years
	0,357652	-0,157153	0,139581	Life expec.
	-0,370114	0,161751	-0,110960	Infant mort.
	0,387353	-0,033306	0,009817	Phones
	0,309539	0,029743	0,052751	Internet
	0,379927	0,084639	0,017924	GDP $/capita
	-0,333591	-0,092654	-0,061686	% GDP agri
	0,272296	0,122392	-0,554596	% GDP ind
	0,019198	-0,065779	0,790732	% GDP ser

In our example, since we have placed three columns, the coefficients of the first three components have been saved. The representation of the third component with respect to the first looks as follows:

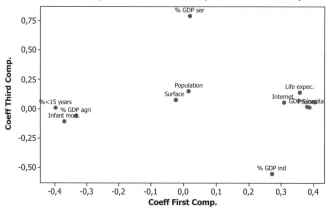

If you want the lines that connect the points with the origin to appear, copy the columns that contain the coordinates of the points and insert cells above each pair of data (select the pair and click on the Insert cells button) from the second row:

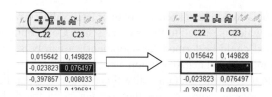

Once all cells have been inserted, replace the asterisks with zeros. Right-click on the figure, **Add > Calculated Line,** and in **Y Column** place the column that contains the values of the third component with the zeros inserted. In **X Column** do the same for the first column.

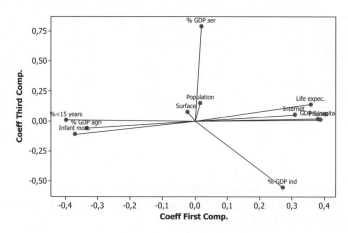

24.3   Cluster Analysis for Observations

Continue working with file LATIN_AMERICA.MTW. Now, the aim is to divide the countries in groups (clusters) of similar characteristics according to the information available.

Stat > Multivariate > Cluster Observations

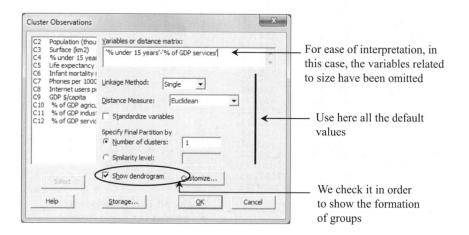

For ease of interpretation, in this case, the variables related to size have been omitted

Use here all the default values

We check it in order to show the formation of groups

Take a look at the output in the Session window (blank lines have been added in order to insert comments).

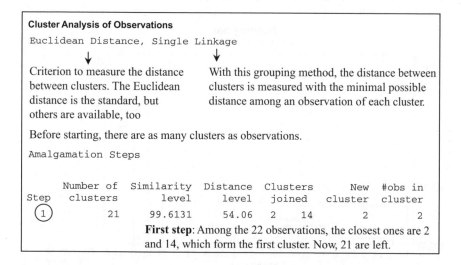

Cluster Analysis of Observations

Euclidean Distance, Single Linkage

Criterion to measure the distance between clusters. The Euclidean distance is the standard, but others are available, too

With this grouping method, the distance between clusters is measured with the minimal possible distance among an observation of each cluster.

Before starting, there are as many clusters as observations.

Amalgamation Steps

Step	Number of clusters	Similarity level	Distance level	Clusters joined		New cluster	#obs in cluster
1	21	99.6131	54.06	2	14	2	2

First step: Among the 22 observations, the closest ones are 2 and 14, which form the first cluster. Now, 21 are left.

| (2) | 20 | 99.4939 | 70.73 | 7 | 12 | 7 | 2 |

Second step: Among the 21 clusters we have, the closest ones are those corresponding to observations 7 and 12.

| (3) | 19 | 99.2755 | 101.25 | 2 | 9 | 2 | 3 |

Third step: Among the 20 clusters left, the closest ones are the first (with observations 2 and 14) and observation 9. This new cluster has now three observations.

4	18	99.2675	102.37	2	5	2	4
5	17	98.9909	141.02	8	18	8	2
6	16	98.9137	151.81	2	8	2	6
7	15	98.7540	174.12	3	16	3	2
8	14	98.7458	175.28	2	11	2	7
9	13	98.1957	252.15	6	15	6	2
10	12	97.9917	280.66	3	4	3	3
11	11	97.9498	286.51	2	6	2	9
12	10	97.2457	384.91	2	7	2	11
13	9	96.6741	464.79	13	17	13	2
14	8	95.7750	590.44	1	2	1	12
15	7	95.4151	640.73	1	3	1	15
16	6	94.7709	730.75	1	13	1	17
17	5	93.5426	902.41	1	20	1	18
18	4	87.1791	1791.70	19	22	19	2
19	3	85.3070	2053.32	10	19	10	3
20	2	84.7016	2137.93	10	21	10	(4)
21	1	81.2502	2620.26	1	10	1	(22)

Finally, all 22 observations are put together in a single cluster.

The graph shows the sequence of the clusters formation. The first one was that of observations 2 and 14 since these have the highest degree of similarity (least distance using the selected criterion). The second is the one of observations 7 and 12. The third one consists of the first cluster and observation 9, etc.

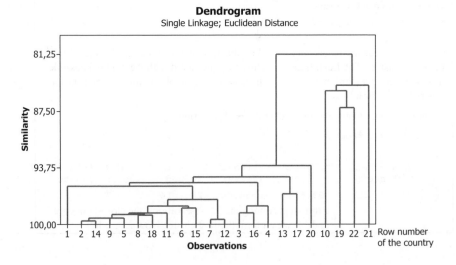

Dendrogram
Single Linkage; Euclidean Distance

Some changes with respect to the default values:

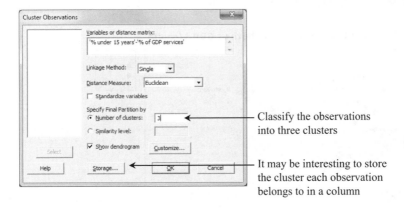

Classify the observations into three clusters

It may be interesting to store the cluster each observation belongs to in a column

Dendrogram
Single Linkage; Euclidean Distance

The 3 clusters appear with lines of different color.

24.3.1 Storage

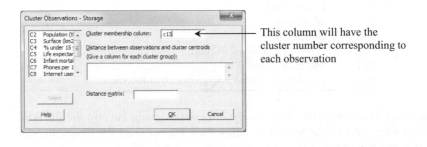

This column will have the cluster number corresponding to each observation

Having identified to which cluster each observation belongs allows the construction of scatterplots identifying the clusters.

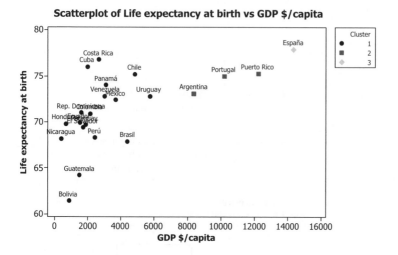

24.4 Cluster Analysis for Variables

Consider again the file CARS.MTW, already used in Chapter 3, that contains the following characteristics of a total of 236 cars:

Columna	Contenido
C1	Maker
C2	Model
C3	Price (in Euros)
C4	Number of cylinders
C5	Displacement (cc)
C6	Power (CV)
C7	Length (cm)
C8	Width (cm)
C9	Height (cm)
C10	Consumption (liters/100 km)
C11	Maximum speed (km/h)
C12	Acceleration (seconds to go from 0 to 100 km/h)

The technique is similar to the cluster analysis for observations, but now the aim is to group variables. This can be helpful to reduce their number grouping them into new ones, or just to have a list of the available 'families' of the variables.

Stat > Multivariate > Cluster Variables

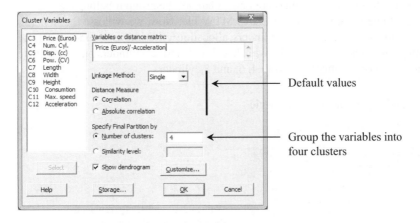

Default values

Group the variables into four clusters

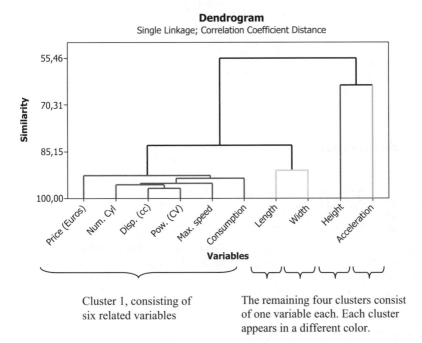

Cluster 1, consisting of six related variables

The remaining four clusters consist of one variable each. Each cluster appears in a different color.

24.5 Discriminant Analysis

The aim of cluster analysis is to group observations, or variables, into initially unknown groups of similar elements. Unlike that, discriminant analysis is applied if the group each observation belongs to is known and one is interested in knowing how the available variables affect the classification, in order to be able to classify a new observation whose variable values are known but not the group it belongs to.

We continue working with the CARS.MTW file. Now, we will use the first 150 cars to establish the number of cylinders allocation criterion as a function of the remaining variables. There are cars with 3, 4, 5, 6, 8 and 12 cylinders but to facilitate the analysis consider only those with 4, 6, and 8 cylinders.

Data > Code > Numeric to Numeric

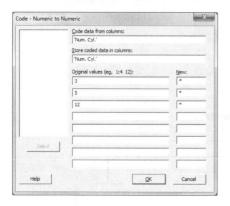

Values 2, 5, and 12 are replaced by * (*missing* value) in the column of the number of cylinders

Data > Subset worksheet

Create a new worksheet with only the first 150 rows

Using this new worksheet, with the cars from rows 1 to 150, perform a discriminant analysis.

Stat > Multivariate > Discriminant Analysis

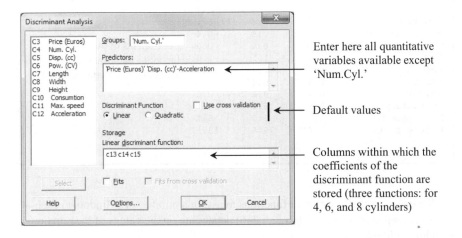

Enter here all quantitative variables available except 'Num.Cyl.'

Default values

Columns within which the coefficients of the discriminant function are stored (three functions: for 4, 6, and 8 cylinders)

In addition to the descriptive information on the data, in the Session Window a summary of the allocations made, the discriminant function for each group and a summary of the errors done appear.

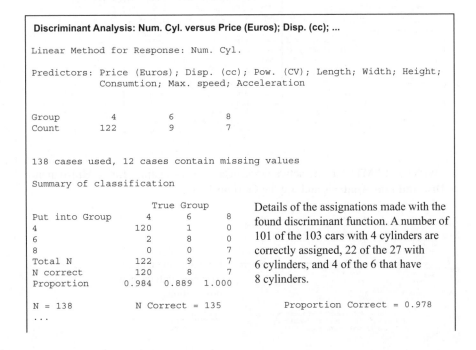

Discriminant Analysis: Num. Cyl. versus Price (Euros); Disp. (cc); ...

Linear Method for Response: Num. Cyl.

Predictors: Price (Euros); Disp. (cc); Pow. (CV); Length; Width; Height;
            Consumtion; Max. speed; Acceleration

Group	4	6	8
Count	122	9	7

138 cases used, 12 cases contain missing values

Summary of classification

	True Group		
Put into Group	4	6	8
4	120	1	0
6	2	8	0
8	0	0	7
Total N	122	9	7
N correct	120	8	7
Proportion	0.984	0.889	1.000

N = 138   N Correct = 135   Proportion Correct = 0.978
. . .

Details of the assignations made with the found discriminant function. A number of 101 of the 103 cars with 4 cylinders are correctly assigned, 22 of the 27 with 6 cylinders, and 4 of the 6 that have 8 cylinders.

```
Linear Discriminant Function for Groups

                      4         6         8
Constant        -1091.4   -1116.6   -1196.5
Price (Euros)      -0.0      -0.0      -0.0
Disp. (cc)          0.0       0.0       0.1      Coefficients of the discriminant
Pow. (CV)          -1.6      -1.5      -1.1      functions per group. These values
Length             -0.1      -0.1      -0.1      are stored in columns C15-C17.
Width               0.4       0.4       0.5
Height              0.2       0.2       0.1
Consumtion          9.1       9.1      10.8
Max. speed          5.1       5.2       5.2
Acceleration       37.6      39.5      43.5

...
```

We can use the discriminant functions to assign a number of cylinders to each of the remaining cars (rows 151 through 236) depending on the values of the remaining variables defined.

First, copy rows 151–236 of file CARS.MTW and paste them into the same worksheet, from column C16.

Already existing columns ◄───┼───► New columns pasted (rows 151–236 of the complete worksheet CARS.MTW)

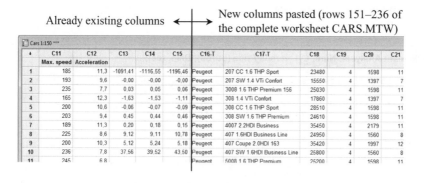

	C11	C12	C13	C14	C15	C16-T	C17-T	C18	C19	C20	C21
	Max. speed	Acceleration									
1	185	11,3	-1091.41	-1116.55	-1196,46	Peugeot	207 CC 1.6 THP Sport	23480	4	1598	11
2	193	9,6	-0,00	-0,00	-0,00	Peugeot	207 SW 1.4 VTi Confort	15550	4	1397	7
3	235	7,7	0,03	0.05	0,06	Peugeot	3008 1.6 THP Premium 156	25030	4	1598	11
4	165	12,3	-1,63	-1,53	-1,11	Peugeot	308 1.4 VTi Confort	17860	4	1397	7
5	200	10,6	-0,06	-0,07	-0,09	Peugeot	308 CC 1.6 THP Sport	28510	4	1598	11
6	203	9,4	0.45	0,44	0,46	Peugeot	308 SW 1.6 THP Premium	24610	4	1598	11
7	189	11,3	0.20	0,18	0,15	Peugeot	4007 2.2HDI Business	35450	4	2179	11
8	225	8,6	9.12	9,11	10,78	Peugeot	407 1.6HDI Business Line	24950	4	1560	8
9	200	10,3	5.12	5,24	5,18	Peugeot	407 Coupe 2.0HDI 163	35420	4	1997	12
10	236	7,8	37.56	39.52	43,50	Peugeot	407 SW 1.6HDI Business Line	25800	4	1560	8
11	245	6.8				Peugeot	5008 1.6 THP Premium	25200	4	1598	11

With CARS.MTW as the active worksheet, execute again: **Stat > Multivariate > Discriminant Analysis** and use the **Options** button:

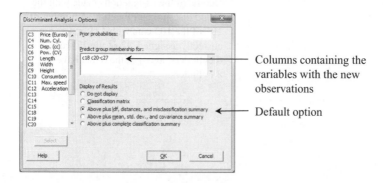

Columns containing the variables with the new observations

Default option

Of the 86 cars we have added, all those with 4 cylinders are classified correctly as such. The only mistake made consists in assigning 8 cylinders to a car that actually has 6. More cars are correctly classified among those with 4 cylinders because there are many more cars of that type in the sample. Hence, the coefficients of the discriminant function are estimated more precisely.

25

Part Five: Case Studies
Regression and Multivariate Analysis

25.1   Tree

To determine the amount of timber that can be extracted from a forest area, equations or tables are often used to estimate the tree's volume as a function of its diameter and/or its height. The file TREE.MTW contains information on 31 trees of a certain type. The data is from B.F. Ryan, B.L. Joiner and T.A. Ryan, Jr, *Minitab Handbook, 2nd edition* (© 1992 Brookes/Coles, a part of Cengage Learning, Inc. reproduced by permission. www.cengage.com./permissions). The file is organized as follows:

Column	Content
C1	Tree diameter at a certain distance from the ground (in cm)
C2	Tree height (in m)
C3	Volume of useable timber (in dm^3)

The aim is to find an equation to determine the timber volume of this type of tree as a function of the diameter, height, or both.

Industrial Statistics with Minitab, First Edition. Pere Grima Cintas, Lluís Marco-Almagro and Xavier Tort-Martorell Llabrés.
© 2012 John Wiley & Sons, Ltd. Published 2012 by John Wiley & Sons, Ltd.

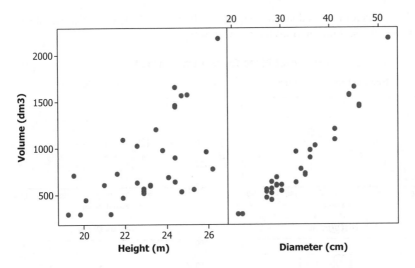

The bivariate plots (**Graph > Matrix Plot > Each Y versus each X: Simple**) clearly show that the best model with only one regressor is the one that explains the volume as a function of the diameter. In the following, this model is studied in detail:

Stat > Regression > Regression

```
The regression equation is
Volume (dm3) = - 1041 + 56.3 Diameter (cm)

Predictor             Coef  SE Coef        T      P
Constant          -1040.52    99.81   -10.42  0.000
Diameter (cm)       56.320     2.889    19.49  0.000

S = 126.045   R-Sq = 92.9%   R-Sq(adj) = 92.7%

Analysis of Variance

Source           DF        SS       MS       F      P
Regression        1   6036609  6036609  379.96  0.000
Residual Error   29    460733    15887
Total            30   6497341

Unusual Observations

      Diameter   Volume
Obs       (cm)    (dm3)      Fit   SE Fit   Residual   St Resid
31        52.0   2180.0   1888.1     57.7      291.9     2.60RX

R denotes an observation with a large standardized residual.
X denotes an observation whose X value gives it large leverage.
```

The plots for the residual analysis (in the dialog box **Regression**, press button **Graphs** and choose **Four in one**) look as follows:

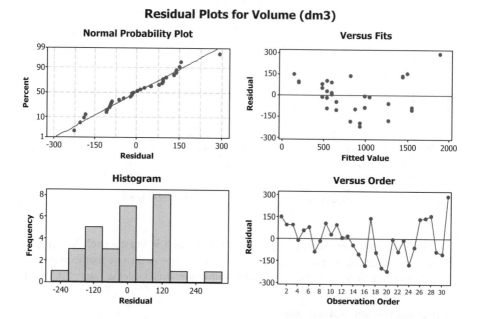

Residual Plots for Volume (dm3)

The most notable issue is that in the second graph, residuals versus fitted values (**Versus Fits**), the points are distributed approximately in a quadratic shape, which suggests that introducing the square of the regressor variable could improve the model. This can be achieved as follows.

Stat > Regression > Fitted Line Plot

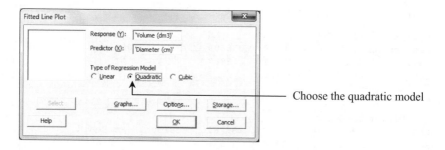

Choose the quadratic model

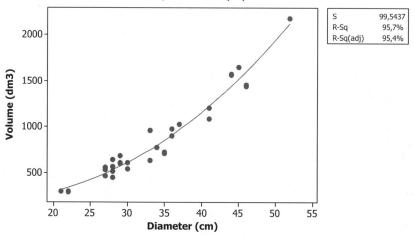

Fitted Line Plot
Volume (dm3) = 359,2 - 26,65 Diameter (cm)
+ 1,166 Diameter (cm)**2

S	99,5437
R-Sq	95,7%
R-Sq(adj)	95,4%

Indeed, a good fit is obtained. Both the value of R^2 (which has increased) and the standard deviation (which is lower now) of the residuals (S) have improved.

To get the list of singular points, select: **Stat > Regression > Regression.**

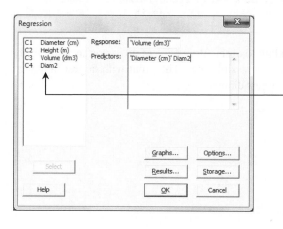

Previously, a column containing the square of the diameter has been created

```
The regression equation is
Volume (dm3) = 359 - 26.7 Diameter (cm) + 1.17 Diam2

Predictor          Coef  SE Coef       T       P
Constant          359.2    334.9    1.07   0.293
Diameter (cm)    -26.65    19.43   -1.37   0.181
Diam2            1.1663   0.2712    4.30   0.000

S = 99.5437    R-Sq = 95.7%    R-Sq(adj) = 95.4%

Analysis of Variance

Source            DF       SS       MS       F       P
Regression         2  6219891  3109946  313.85   0.000
Residual Error    28   277450     9909
Total             30  6497341

Source            DF   Seq SS
Diameter (cm)      1  6036609
Diam2              1   183282

Unusual Observations

      Diameter  Volume
Obs       (cm)   (dm3)      Fit  SE Fit  Residual  St Resid
17        33.0   957.0    749.8    23.9     207.2      2.14R
31        52.0  2180.0   2127.1    71.8      52.9      0.77 X

R denotes an observation with a large standardized residual.
X denotes an observation whose X value gives it large leverage.
```

The large residual corresponding to observation 17 is of no importance (it is only 2.14 standard deviations from the regression line). The point with largest influence corresponds to the tree with the largest diameter.

Nothing remarkable is seen in the residuals analysis plots.

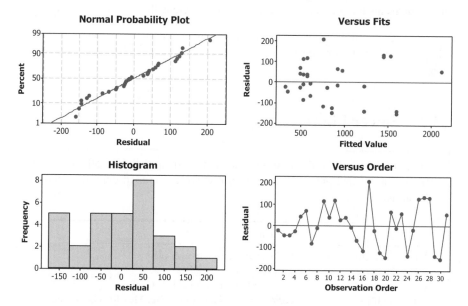

However, since it is known that the volume of a cylinder is the product of the base area and its height, a new variable is created as the squared diameter times the height. With that new variable, a better fit is obtained.

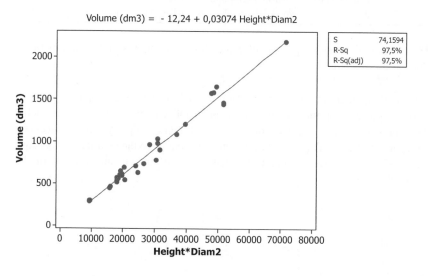

Again, both the coefficient of determination (R^2) and the standard deviation of the residuals have improved with respect to the previous fit.

 Use your prior knowledge about the phenomenon under study in order to find the best models.

25.2   Power Plant

The file POWER_PLANT.MTW contains data corresponding to 50 days of operation of a thermal power station. Each row corresponds to daily data and the columns are:

Col.	Name	Content
C1	Yield	Yield of the thermal power plant
C2	Power	Average power
C3	Fuel	Combustible used (0: Fuel, 1: Gas)
C4	FF	Form factor of the power curve. Measures the variation of the power throughout the day.
C5	Steam Temp.	Live steam temperature
C6	Air Temp.	Air temperature (environment)
C7	Seawater Temp.	Seawater temperature (cold source)
C8	Day	Weekday (1: Monday, 2: Tuesday, ...)

The aim consists in building a model that explains the plant's yield as a function of the variables available.

First, it should be clear that the variable Day, as such (qualitative variable with more than two categories), cannot be a candidate to be included in the model. However, what can be done is to encode it as 0=working day (Monday to Friday) and 1=weekend (Saturday and Sunday). This distinction seems reasonable to the technicians, because on weekends there is less demand and that could affect the performance of the power plant.

Data > Code > Numeric to a Numeric

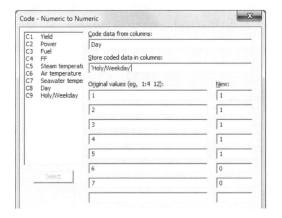

With the new variable created, we carry out an exploratory analysis of the data.

Graph > Matrix Plot: Simple

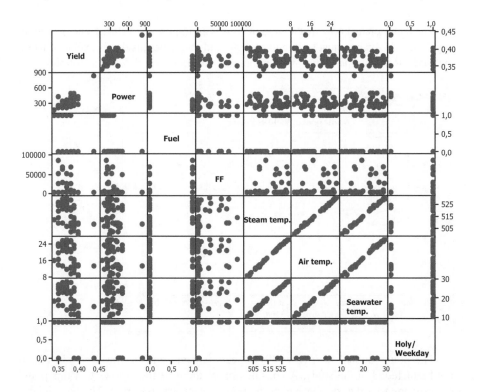

As there are many variables, the graph is not too clear. Nonetheless, in the scatterplot of the yield versus the power, a single point far from the others can be seen. It shows a day with both high yield and high power. To have a closer look at this graph, choose: **Graph > Scatterplot: Simple**

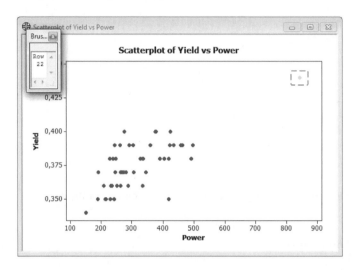

The technicians know that a yield of 0.44 is an impossible value. Hence, without any doubt, this value must be due to an error and therefore this day is excluded from the study. Redoing the scatterplot without that point, the graph looks as follows:

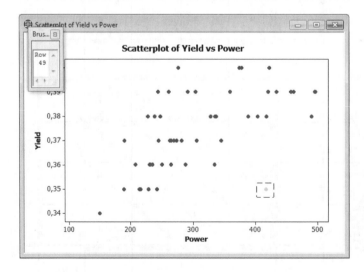

The point marked now (row 49) appears isolated and maybe it should better not be included in the study. A consultation of the data files shows that this point corresponds to a day during which the power plant was restarted. The technicians consider that this day should not be included in the study because there are only few such days with a restart of the power plant and because the performance during these days is exceptionally low. The model should not be based on such days. Hence, this data point is also deleted.

Deleting the value of the response variable (i.e. considering it as a missing value and substituting it by an asterisk) is equivalent to eliminating all coordinates of the point (in our case, eliminating the data of the whole row).

In case one wants to delete the value of a regressor variable, this should be marked as a missing, but all other coordinates of the point should be kept. If the variable with the missing value is included in the model, the data point is completely ignored; however, if the variable is not included, no data are lost.

Before starting to look for the best model we add some variables, transformations of the original ones, which may improve or better explain the response variable. The transformations considered reasonable are:

- The power to the square: The scatterplot of yield versus power shows a relation that can be better described by means of a quadratic curve than by a straight line.

- The inverse of the power: This type of nonlinear relationship can also be modelled through the inverse of the power. In addition, our knowledge of the yield's formula makes us think that the inverse of the power is a reasonable variable.

- The logarithm of the form factor: The form factor shows some values grouped near zero. Its dotplot (**Graph > Individual value plots > One Y: Simple**) looks as follows:

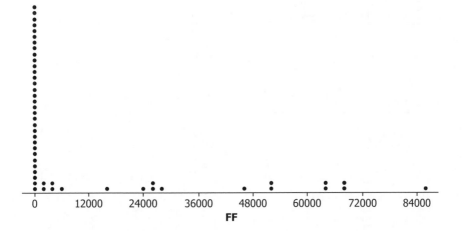

We know that, in general, information is better extracted from variables with such a behaviour if a logarithmic transformation is applied. After the transformation, the new dotplot has the following aspect:

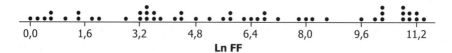

With the transformed variables, the worksheet looks as follows:

Response		Possible original regressor variables				Will not be considered		Possible transformed regressor variables		

POWER_PLANT.MTW ***

↓	C1	C2	C3	C4	C5	C6	C7	C8	C9	C10	C11	C12
	Yield	Power	Fuel	FF	Steam temp.	Air temp.	Seawater temp.	Day	Holy/Weekday	Power2	1/Power	Ln FF
5	0.36	287.8	1	62.9	523.9	22.8	26.0	5	1	82829	0.0034746	4.1415
6	0.38	336.8	1	68819.0	509.2	12.8	16.0	5	1	113434	0.0029691	11.1392
7	0.35	214.1	1	51708.5	523.9	22.8	25.3	4	1	45839	0.0046707	10.8534
8	0.37	345.2	1	15674.6	520.2	20.6	23.1	3	1	119163	0.0028969	9.6598
9	0.35	229.0	1	25090.9	519.0	19.8	22.8	2	1	52441	0.0043668	10.1303
10	0.35	216.4	1	86997.6	513.8	16.5	19.1	4	1	46829	0.0046211	11.3736
11	0.36	333.7	1	274.4	523.6	22.3	25.8	1	1	111356	0.0029967	5.6146

We generate all possible models as follows.

Stat > Regression > Best Subsets

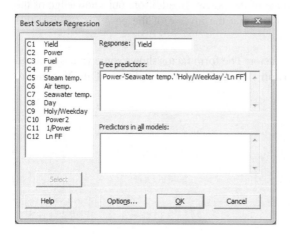

With all the default options, we obtain:

```
Response is Yield
49 cases used, 1 cases contain missing values
                                              S H
                                           S  e o
                                           t  a l
                                           e At y
                                           a ie /     1
                                           m rr W P   /
                          P                t tte o P  L
                          o  F             e eek w o  n
                          w  u             m mmd e w
              Mallows     e  e F p p p a r e F
Vars  R-Sq  R-Sq(adj)  Cp        S      r  l F . . . y 2 r F
  1   60.4    59.5      97.8   0.0099946 X
  1   51.1    50.0     131.0   0.011103                        X
  2   75.1    74.0      47.1   0.0080112 X                     X
  2   72.6    71.4      56.1   0.0084063               X       X
  3   83.0    81.8      20.8   0.0066924 X             X       X
  3   82.9    81.7      21.4   0.0067234 X  X                  X
  4   87.1    85.9       8.1   0.0058957 X                X    X X
  4   87.1    85.9       8.2   0.0059020 X             X       X X
  5   89.0    87.7       3.2   0.0055029 X X               X   X X  <---
  5   89.0    87.7       3.4   0.0055123 X X           X       X X
  6   89.4    87.8       4.0   0.0054796 X X           X     X X X
  6   89.3    87.8       4.2   0.0054952 X X               X X X X
  7   89.5    87.7       5.4   0.0055060 X X       X X       X X X
  7   89.4    87.5       5.9   0.0055430 X X X     X         X X X
  8   89.6    87.5       7.2   0.0055632 X X     X X X       X X X
  8   89.6    87.4       7.3   0.0055672 X X X X X X         X X X
  9   89.6    87.2       9.1   0.0056234 X X X X X X         X X X
  9   89.6    87.2       9.1   0.0056291 X X X X X         X X X X
 10   89.7    86.9      11.0   0.0056933 X X X X X X X X X X
```

The most interesting model is indicated with an arrow (low values of Cp and S, high values of R-Sq(adj)). We will have a closer look at it.

Stat > Regression > Regression

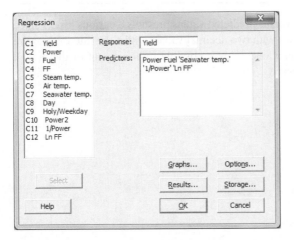

In **Graphs,** we select option **Four in one** in order to produce the residual plots.

```
The regression equation is
Yield = 0.476 - 0.000078 Power - 0.0122 Fuel - 0.000808
         Seawater temp. - 14.1 1/Power - 0.000959 Ln FF
48 cases used, 2 cases contain missing values

Predictor               Coef    SE Coef      T      P
Constant             0.47636    0.01806   26.38  0.000
Power            -0.00007802  0.00002876   -2.71  0.010
Fuel               -0.012194    0.002103   -5.80  0.000
Seawater temp.     -0.0008076   0.0001559   -5.18  0.000
1/Power              -14.114       2.554   -5.53  0.000
Ln FF              -0.0009585   0.0002325   -4.12  0.000

S = 0.00550286    R-Sq = 89.0%    R-Sq(adj) = 87.7%

Analysis of Variance

Source          DF        SS         MS      F      P
Regression       5  0.0103261  0.0020652  68.20  0.000
Residual Error  42  0.0012718  0.0000303
Total           47  0.0115979

Unusual Observations

Obs  Power     Yield        Fit   SE Fit  Residual St Resid
4      150  0.340000   0.342651 0.004282 -0.002651    -0.77X
15     263  0.370000   0.356510 0.001672  0.013490     2.57R
24     335  0.380000   0.397380 0.001903 -0.017380    -3.37R

R denotes an observation with a large standardized residual.
X denotes an observation whose X value gives it large leverage.
```

The graph of the residuals versus predicted values presents a curious aspect since the values of the yield are shown with two decimal values whilst the fitted values are calculated with more precision. Concerning a possible improvement of the model, nothing really relevant is observed.

Residual Plots for Yield

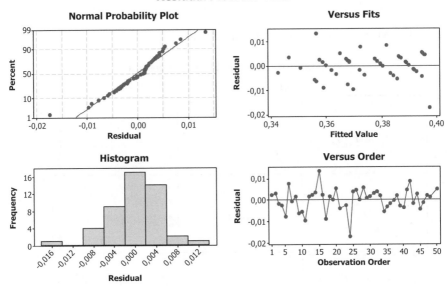

Before finally accepting the model as the final model, it has been checked that nothing exceptional happened the days that appeared as unusual observations (4, 15 and 24). There is no reason to exclude these data from the study and, hence, we accept the model.

25.3 Wear

To study the possible relationship between the wear and tear of a cutting machine and both its rotation speed and the material it was made of, the data of file WEAR.MTW are provided:

Col.	Content
C1	Wear and tear after 8 hours of operation (in tenths of a mm)
C2	Rotation speed (in rpm)
C3	Type of steel (encoded as 0 and 1)

The aim is to find an equation that explains the wear and tear as a function of both variables available.

As always, we start with an exploratory analysis of the data:

Graph > Matrix Plot > Matrix of plots: Simple

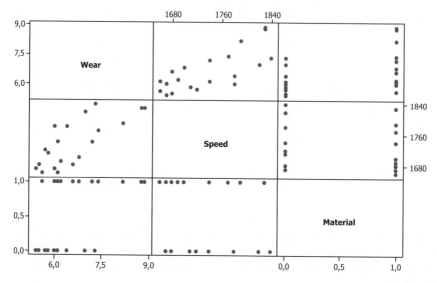

Looking at the **Matrix Plot**, we can rule out the presence of anomalous values. Regarding the relations among the variables, we see that a higher speed is associated with a major wear, and that material 1 wears out more than material 0.

In this case, it is also a good idea to draw a scatterplot of wear versus speed stratified according to the material.

Graph > Scatterplot: With Groups

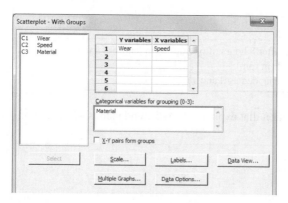

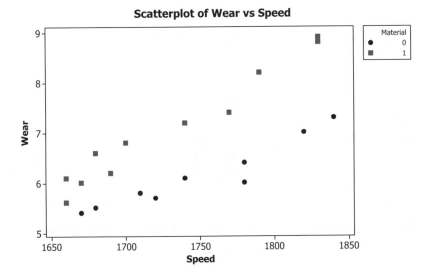

We can clearly see the relation between wear and speed as well as that material 1 wears out more than material 0. We compute the equation of the regression model:

Stat > Regression > Regression

```
The regression equation is
Wear = - 19.0 + 0.0144 Speed + 1.22 Material

Predictor       Coef     SE Coef       T       P
Constant      -18.973      1.975    -9.61   0.000
Speed        0.014356   0.001128    12.73   0.000
Material       1.2236      0.1382    8.85   0.000

S = 0.303451   R-Sq = 92.5%   R-Sq(adj) = 91.6%
```

Since the type of material is a categorical variable with values 0 and 1, we actually have two models which are obtained by substituting these values in the equation:

Model for the type 0 material (Variable Material = 0):
Wear and tear = −19.00 + 0.0144 Velocity

Model for the type 1 material (Variable Material = 1):
Wear and tear = −17.78 + 0.0144 Velocity

We can represent the scatterplot with the regression lines corresponding to each material. To do so, we calculate the values of the wear for the speeds of 1650 and 1850 rpm (coordinates of the origin and the end of each line).

```
MTB > let c5(1)=-18,973+0,0144*1650
MTB > let c5(2)=-18,973+0,0144*1850
MTB > let c6(1)=-17,753+0,0144*1650
MTB > let c6(2)=-17,753+0,0144*1850
MTB >
```

It may also be calculated with the **Calculator** or calculated apart.

WEAR.MTW ***

↓	C1	C2	C3	C4	C5	C6	
	Wear	Speed	Material	X	Y0	Y1	
1	5,5	1680	0	1650	4,787	6,007	
2	8,2	1790	1	1850	7,667	8,887	
3	6,0	1670	1				
4	5,7	1720	0				

On the scatterplot, make a right click on the button: **Add > Calculated Line**

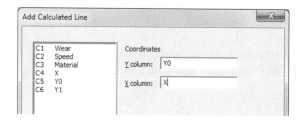

To add another line, the process has to be repeated placing column Y1 and X.

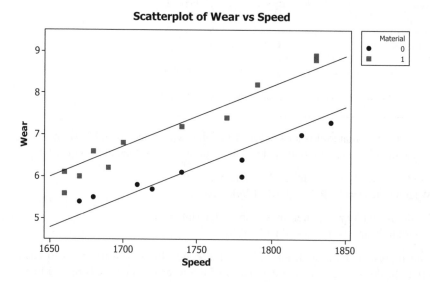

Naturally, the slope of both lines is the same since the same coefficient for the speed was used. However, it does not seem that the speed affects equally both materials, i.e. in a more adequate model, the slope should not be the same for both materials.

The way to allow for two different regression line slopes is to include the speed × material interaction in the model (the effect of the speed depends on the material). Creating this new variable (the product of speed × material) we obtain:

```
The regression equation is

Wear = - 12.1 + 0.0104 Speed - 10.1 Material + 0.00650 Speed*Mat

Predictor        Coef    SE Coef        T       P
Constant      -12.114      2.359    -5.14   0.000
Speed        0.010434   0.001348     7.74   0.000
Material      -10.088      3.023    -3.34   0.004
Speed*Mat    0.006497   0.001735     3.74   0.002

S = 0.228361   R-Sq = 96.0%   R-Sq(adj) = 95.2%
```

The goodness of fit measures of the model have improved and the models are now:

Model for the material of type 0 (Variable Material = 0):
Wear and tear $= -12.1 + 0,0104$ Velocity

Model for the material of type 1 (Variable Material = 1):
Wear and tear $= -22.2 + 0,0169$ Velocity

As done before, we represent both regression lines in a scatterplot:

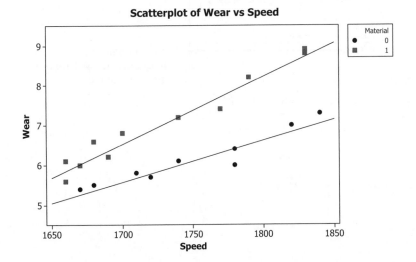

Now, the fit looks very correct and the residual analysis does not suggest that the model could be improved. Hence, we accept it as the final model.

25.4   TV Failure

A certain component used in consumer electronic devices loosens and deteriorates over time for unknown reasons. The data of 113 components installed three years before are available and for each of these, several electronic and mechanical characteristics that were thought to be related with premature failure were recorded.

The data are stored in file TVFAILURE.MTV. Columns C1 to C5 contain the values of these characteristics and C6 indicates deterioration (1) or not (0) after three years of operation.

We would like to find a criterion to predict a possible failure within the first three years of operation.

Here, we are dealing with a typical case for which discriminant analysis can be useful.

Stat > Multivariate > Discriminant Analysis

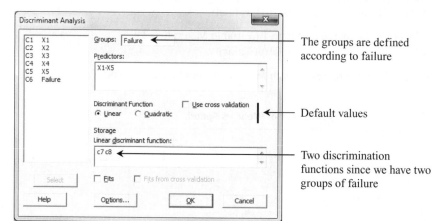

The groups are defined according to failure

Default values

Two discrimination functions since we have two groups of failure

Discriminant analysis: Fal vs. X1; X2; X3; X4; X5

```
Linear Method for Response: Failure

Predictors: X1; X2; X3; X4; X5

Group           0            1
Count          83           30
```

Summary of classification

```
                 True Group
Put into Group      0       1
0                  78       0
1                   5      30
Total N            83      30
N correct          78      30
Proportion      0.940   1.000
```

The sample consists of 30 failures and 83 nonfailures. All 30 failures are correctly classified, whereas among the 83 nonfailures, 78 are classified correctly and 5 are erroneously classified as failures.

```
N = 113           N Correct = 108          Proportion Correct = 0.956
```

Squared Distance Between Groups

```
          0          1
0   0.00000    7.16712
1   7.16712    0.00000
```

Linear Discriminant Function for Groups

```
                0          1
Constant  -548.01    -508.97
X1           1.83       4.62
X2         -19.52     -16.74
X3          28.56      25.11
X4          13.08      10.36
X5          30.96      29.23
```

The coefficients of the discrimination functions. Given a new product, the group is assigned according to the discrimination function that gives a larger value.

Summary of Misclassified Observations

Observation	True Group	Pred Group	Group	Squared Distance	Probability
29**	0	1	0	4.770	0.332
			1	3.375	0.668
36**	0	1	0	10.826	0.134
			1	7.099	0.866
62**	0	1	0	9.383	0.092
			1	4.809	0.908
65**	0	1	0	10.219	0.163
			1	6.946	0.837
72**	0	1	0	6.137	0.419
			1	5.482	0.581

Details on the wrongly classified observations

As an example of application of the discrimination functions, suppose we have 14 components on which characteristics X1–X5 were measured and we wish to predict whether they will suffer a failure or not in the following 3 years (data of file TVFAILURE example.MTW).

TVFAILUREexample.MTW ***

↓	C1	C2	C3	C4	C5	C6	C7	C8	C9	C10	C11	C
		X1	X2	X3	X4	X5	FD1	FD2	Result FD1	Result FD2	Failure	
1	1	8,88	0,382	2,25	1,25	31,43	-548,006	-508,972	514,150	513,563	0	
2	1	10,98	0,604	1,69	0,69	33,27	1,833	4,623	547,711	553,824	1	
3	1	8,09	0,033	1,87	0,87	32,29	-19,518	-16,742	530,458	527,560	0	
4	1	8,73	-0,142	2,18	1,18	30,26	28,564	25,112	485,219	485,172	0	
5	1	9,16	0,618	1,60	0,60	33,33	13,083	10,357	541,937	543,510	1	
6	1	8,87	0,199	2,01	1,01	30,70	30,956	29,227	485,566	487,111	1	
7	1	6,90	0,363	1,92	0,92	31,50			499,323	495,069	0	
8	1	11,09	0,426	1,37	0,37	30,29			445,571	458,674	1	
9	1	9,72	0,382	2,07	1,07	31,91			523,227	525,228	1	
10	1	8,42	0,795	2,25	1,25	32,42			536,051	533,610	0	
11	1	7,78	0,364	1,39	0,39	32,52			510,416	510,135	0	
12	1	8,82	-0,029	1,80	0,80	33,27			560,388	558,068	0	
13	1	9,36	0,615	1,92	0,92	32,80			539,376	540,427	1	
14	1	10,79	0,241	2,21	1,21	31,92			534,291	537,985	1	
15												

Columns of ones to facilitate matrix operations

Coefficients of the discriminations functions found

If the value in C10 is larger than the one in C9, a failure is predicted:

```
MTB > let c11=c10>c9
```

Values of the new components

Results of applying both discrimination functions to each component. The most comfortable way is to use matrix operations by either using the menus or writing directly:

```
MTB > copy c1-c6 m1

MTB > copy c7 m2
MTB > copy c8 m3
MTB > multi m1 m2 m4
MTB > multi m1 m3 m5
MTB > copy m4 c9
MTB > copy m5 c10
```

Part Six

EXPERIMENTAL DESIGN AND RELIABILITY

One must try by doing the thing; for though you think you know it, you have no certainty until you try

Sophocles

Imagine a biscuit manufacturing process; simplifying it can be divided into two basic phases: dough preparation and baking. Each involves a large number of variables under our control, from the proportion of each mass component, to temperature, time and moisture during cooking. Design of experiments is a methodology, rather than a tool, aimed at learning how each of these factors affects the cookies' characteristics of interest: color, hardness, degree of crispness, etc.

We are therefore faced with a learning method. It is often surprising how little is known about the behaviour of industrial processes and products. For questions such as: What happens to the hardness of the biscuits if we increase the temperature of the oven? Will it affect some other characteristic? Frequently there are as many answers as technicians consulted. Naturally, the situation gets worse when the question gets more complicated: What happens to the hardness when the oven temperature is increased and the proportion of butter and the cooking time are decreased?

By asking yourself similar questions about processes you know about, you will probably realize that your knowledge about these, in theory, well-known processes is weak. In most cases the answers to the questions will be unknown or vague and not quantified. Rarely will the answer be like, 'if we increase the oven temperature 10 degrees and decrease by 1% the proportion of butter, a biscuit's hardness will increase by 6 units.

There are two ways to obtain precise scientific answers to such questions: through theoretical knowledge – rarely available to the required level of detail – or through experimentation. One could argue that a third way is through experience, but experience is just the accumulation of knowledge based on experimentation – even if conducted in a disorganized and unconscious way – over a lengthy period of time. So, in the end, experimental design is about learning, about gaining knowledge, even if many times the final objective is to optimize a process.

There is wide consensus on the usefulness of experimentation to improve products and processes. The problem is that it is an expensive activity – in terms of time, resources and knowledge. This is so, because when experiments are to be conducted on the production line, they interfere with production plans and schedules, and when they are aimed at product design or improvement they require building prototypes, frequently an expensive or difficult task. However, conducting trials using intuitive strategies is far more expensive. Statistical experimental design methods are designed to maximize the ratio: information-experimental effort.

Minitab offers a complete range of design options, but we only deal with 2^k full or fractional factorial designs. Both, the statistical concepts involved and the Minitab options are similar in other designs. Chapter 26, the first chapter of Part 6, is dedicated to design selection; the next one to the analysis and interpretation of results and a third one to response surface methodology, useful when linear approximations are not sufficient to explain the response behaviour.

Obviously, neither Minitab nor any other statistical package solves some of the key factors in the success of an experiment: which factors to include, at which levels, how to measure the response, etc. That is why it is so important the collaboration between process experts and experimental design specialists.

The last chapter of this part is dedicated to reliability, a very important topic in industrial statistics and in our experience, a frequently misused or forgotten one. Obviously, things like the life expectancy of a product or a piece of equipment, the failure rate after a certain number of hours of operation, the expected warranty costs or the frequency of preventive maintenance operation are of general interest. To tackle those questions it is necessary, as usual, to collect adequate data and to analyze it properly. In general, it is recommended to first conduct a *nonparametric analysis* based only on the data, without assuming any probability distribution; and then find an appropriate distribution – Weibull, lognormal, etc. – and analyze the data again to obtain richer conclusions, by this *parametric analysis*. Minitab facilitates both types of analysis and also provides a very convenient option to identify, if it exists, a probability distribution that fits the data well.

26

Factorial Designs: Creation

26.1 Creation of the Design Matrix

Stat > DOE > Factorial > Create Factorial Design

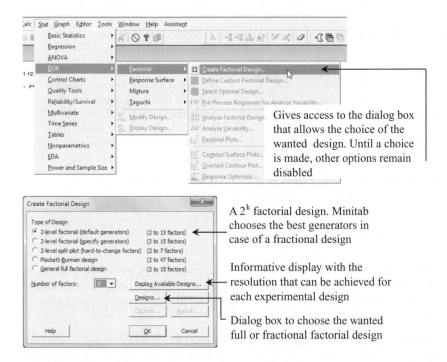

Gives access to the dialog box that allows the choice of the wanted design. Until a choice is made, other options remain disabled

A 2^k factorial design. Minitab chooses the best generators in case of a fractional design

Informative display with the resolution that can be achieved for each experimental design

Dialog box to choose the wanted full or fractional factorial design

Industrial Statistics with Minitab, First Edition. Pere Grima Cintas, Lluís Marco-Almagro and Xavier Tort-Martorell Llabrés.
© 2012 John Wiley & Sons, Ltd. Published 2012 by John Wiley & Sons, Ltd.

26.1.1   Available Designs

Stat > DOE > Factorial > Create Factorial Design > Display Available Designs

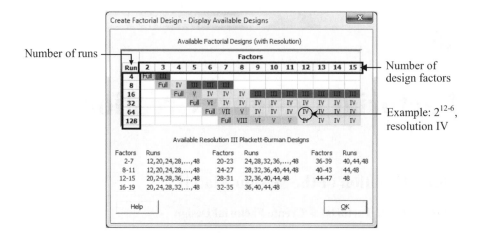

Number of runs

Number of design factors

Example: 2^{12-6}, resolution IV

Stat > DOE > Factorial > Create Factorial Design > Designs

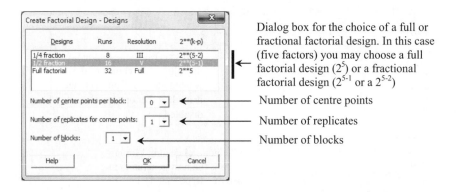

Dialog box for the choice of a full or fractional factorial design. In this case (five factors) you may choose a full factorial design (2^5) or a fractional factorial design (2^{5-1} or a 2^{5-2})

Number of centre points

Number of replicates

Number of blocks

Number of centre points per block (zero by default): If you use centre points you can determine if a model requires quadratic terms, or if instead only linear terms and interactions of order two are needed (for further information, refer to the chapter of response surface).

Number of replicates for corner points (one by default): Making replicas allows you to have an estimate of the standard deviation of the effects and perform tests of significance for each of them.

Number of blocks (one by default): If for some reason you are compelled to perform the experiments in different blocks (for example, in two days) enter here the number of blocks so that each experimental condition is assigned to one of these blocks (in this way, you will be able to interpret the results taking into account the possible effect of the block factor).

 Example 26.1: The welding process performed in a stainless steel component in a production line of exhaust pipes for the automotive industry must be optimized. To achieve this, a 2^3 factorial design is carried out, considering the factors:

	Level −	Level +
A. Gas flow (l/min)	8	12
B. Intensity (A)	230	240
C. Chain speed (m/min)	0.6	1

The response variable is the quality of the components, measured on a scale from zero to 30 (the greater the value, the better the quality).

Use Minitab to design the matrix of the experiment: **Stat > DOE > Factorial > Create Factorial Design**

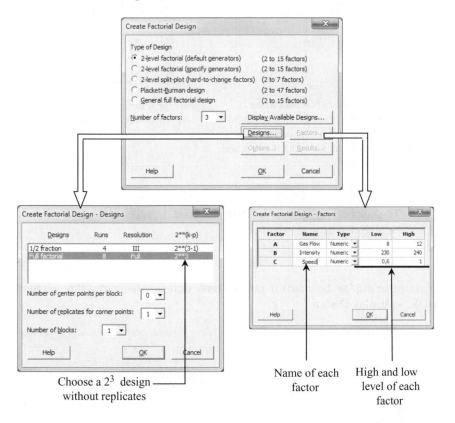

Choose a 2^3 design without replicates

Name of each factor

High and low level of each factor

With button **Options** choose whether to randomize or not the order of experimentation:

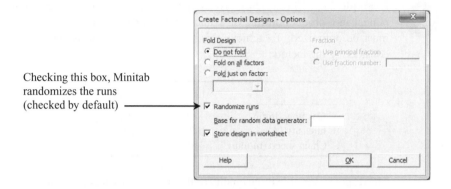

Checking this box, Minitab randomizes the runs (checked by default) ──────

Carry out each run of the experiment in the indicated random order (from the first to the last row). In this case, start producing components with a gas flow of 8 l/min, intensity of 240 A and chain speed of 1 m/min

Register the results of the experiment in the worksheet. In this way, they are ready to be analyzed

The numbers on column **StdOrder** indicate the order in which rows must be placed to be in standard order.

You can either display the matrix in random order or in standard order, through **Stat > DOE > Display Design**

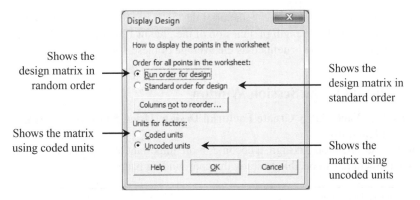

Shows the design matrix in random order

Shows the design matrix in standard order

Shows the matrix using coded units

Shows the matrix using uncoded units

Matrix using coded units

C5	C6	C7	C8
Gas Flow	Intensity	Speed	Y
-1	1	1	17,5
1	1	-1	17,5
1	-1	1	26,0
1	-1	-1	26,5
-1	-1	-1	10,0
-1	1	-1	15,0
1	1	1	20,0
-1	-1	1	11,5

Matrix using uncoded units

C5	C6	C7	C8
Gas Flow	Intensity	Speed	Y
8	240	1,0	17,5
12	240	0,6	17,5
12	230	1,0	26,0
12	230	0,6	26,5
8	230	0,6	10,0
8	240	0,6	15,0
12	240	1,0	20,0
8	230	1,0	11,5

 If you are copying results in the worksheet and they are in standard order, be sure the design matrix in the worksheet is also in standard order (by default it is not).

 Example 26.2: A cava (Spanish champagne) producer carries out an experiment to improve the labelling process. The aim is to increase the adhesion (measured as the force needed to slide the label within the first 30 minutes after it has been glued to the bottle). This is needed to reduce the number of bottles with crooked or decentred labels caused by manipulations done after the labelling process.

The considered variables and corresponding levels are:

		Level −	Level +
A	Type of Glue	X	Y
B	Temperature of Glue	30 °C	40 °C
C	Quantity of Glue	2 gr	3 gr
D	Drying Temperature	80 °C	90 °C
E	Brush Pressure	1 kg	1,5 kg

In a first experiment, a 2^{5-2} factorial design is carried out. In each run, a measure of the strength of adhesion of the glue in 100 bottles is taken; the response variable is defined as the average of these 100 values.

26.1.2   Output in the Session Window

Stat > DOE > Factorial > Create Factorial Design (Using 5 factors, in **Designs** choose a **1/4 fraction**)

Since this is a fractional design, it becomes more important to visualize the output in the session window to be able to study the confounding patterns:

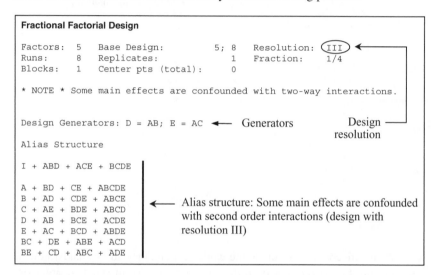

```
Fractional Factorial Design

Factors:  5    Base Design:          5; 8    Resolution:  III
Runs:     8    Replicates:              1    Fraction:    1/4
Blocks:   1    Center pts (total):      0

* NOTE * Some main effects are confounded with two-way interactions.


Design Generators: D = AB; E = AC  ←—— Generators          Design
                                                           resolution
Alias Structure

I + ABD + ACE + BCDE

A + BD + CE + ABCDE
B + AD + CDE + ABCE
C + AE + BDE + ABCD      ←—— Alias structure: Some main effects are confounded
D + AB + BCE + ACDE          with second order interactions (design with
E + AC + BCD + ABDE          resolution III)
BC + DE + ABE + ACD
BE + CD + ABC + ADE
```

Enter the experimental result in the worksheet, as shown below:

↓	C1	C2	C3	C4	C5-T	C6	C7	C8	C9	C10
	StdOrder	RunOrder	CenterPt	Blocks	Type	Temp	Quantity	Drying Temp.	Brush Pres.	Y
1	7	1	1	1	X	40	3	80	1,0	24,5
2	5	2	1	1	X	30	3	90	1,0	25,0
3	1	3	1	1	X	30	2	90	1,5	24,0
4	2	4	1	1	Y	30	2	80	1,0	16,0
5	3	5	1	1	X	40	2	80	1,5	22,5
6	6	6	1	1	Y	30	3	80	1,5	16,0
7	4	7	1	1	Y	40	2	90	1,0	24,5
8	8	8	1	1	Y	40	3	90	1,5	23,5
9										

Experimental result ———↑

 Do not be surprised if most options of menu **DOE > Factorial** appear disabled when you open Minitab. These are activated whenever their use makes sense.

26.2 Design Matrix with Data Already in the Worksheet

If the matrix design and the response are previously entered in the worksheet without using the option **Create Factorial Design**, Minitab must be able to recognize such design matrix prior the statistical analysis. To achieve this, use the option **Define Custom Factorial Design**.

Factors ⟶ 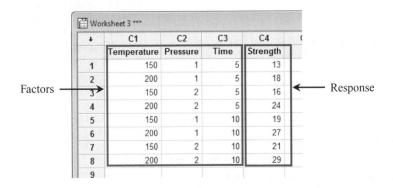 ⟵ Response

Stat > DOE > Factorial > Define Custom Factorial Design

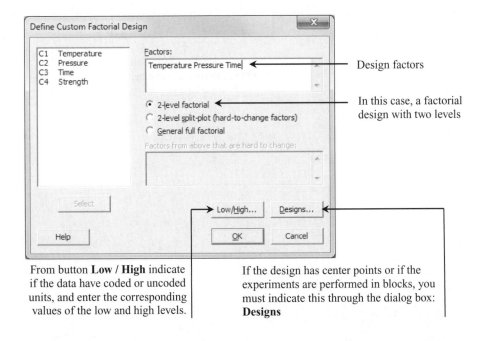

Design factors

In this case, a factorial design with two levels

From button **Low / High** indicate if the data have coded or uncoded units, and enter the corresponding values of the low and high levels.

If the design has center points or if the experiments are performed in blocks, you must indicate this through the dialog box: **Designs**

Stat > DOE > Factorial > Define Custom Factorial Design > Low/High

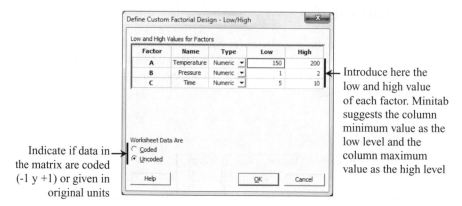

Introduce here the low and high value of each factor. Minitab suggests the column minimum value as the low level and the column maximum value as the high level

Indicate if data in the matrix are coded (-1 y +1) or given in original units

After specifying the wanted design, the worksheet window looks as follows:

	C1	C2	C3	C4	C5	C6	C7	C8
	Temperature	Pressure	Time	Strength	StdOrder	RunOrder	Blocks	CenterPt
1	150	1	5	13	1	1	1	1
2	200	1	5	18	2	2	1	1
3	150	2	5	16	3	3	1	1
4	200	2	5	24	4	4	1	1
5	150	1	10	19	5	5	1	1
6	200	1	10	27	6	6	1	1
7	150	2	10	21	7	7	1	1
8	200	2	10	29	8	8	1	1
9								

Minitab adds these extra columns,
needed to perform the statistical analysis
of the design data.

27

Factorial Designs: Analysis

27.1 Calculating the Effects and Determining the Significant Ones

Stat > DOE > Factorial > Analyze Factorial Design

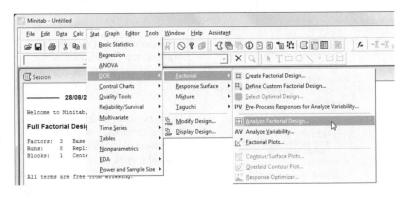

Industrial Statistics with Minitab, First Edition. Pere Grima Cintas, Lluís Marco-Almagro and Xavier Tort-Martorell Llabrés.
© 2012 John Wiley & Sons, Ltd. Published 2012 by John Wiley & Sons, Ltd.

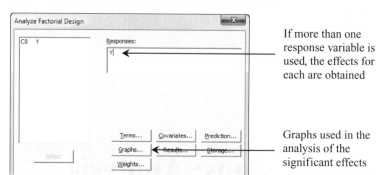

If more than one response variable is used, the effects for each are obtained

Graphs used in the analysis of the significant effects

 This menu option only appears enabled if the worksheet contains a design of experiments matrix. Thus, either the option **Create Factorial Design** or **Define Custom Factorial Design** must have been used previously.

Stat > DOE > Factorial > Analyze Factorial Design > Graphs

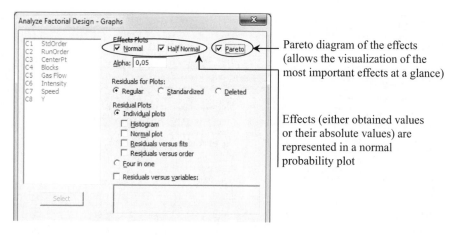

Pareto diagram of the effects (allows the visualization of the most important effects at a glance)

Effects (either obtained values or their absolute values) are represented in a normal probability plot

 Example 27.1: Consider again the first example of Chapter 26 where a design of experiments is carried out to improve the quality of the welding process performed in a stainless steel component, which is used in the production of exhaust pipes. The considered factors are gas flow, intensity and chain speed.

The Minitab worksheet looks as follows:

	C1	C2	C3	C4	C5	C6	C7	C8
	StdOrder	RunOrder	CenterPt	Blocks	Gas Flow	Intensity	Speed	Y
1	2	1	1	1	12	230	0,6	26,5
2	5	2	1	1	8	230	1,0	11,5
3	6	3	1	1	12	230	1,0	26,0
4	3	4	1	1	8	240	0,6	15,0
5	8	5	1	1	12	240	1,0	20,0
6	7	6	1	1	8	240	1,0	17,5
7	4	7	1	1	12	240	0,6	17,5
8	1	8	1	1	8	230	0,6	10,0

If you want to reproduce this example using the option **Create Factorial Design** to generate the design matrix, most probably the runs will be placed in an order different from the one shown. Thus, make sure to place each response value in the corresponding run.

Use **Stat > DOE > Factorial > Analyze Factorial Design**, with column C8 as response, and request (pressing the **Graphs** button) the construction of the three graphical representations of the effects to obtain:

Factorial Fit: Y versus Gas Flow; Intensity; Speed

Estimated Effects and Coefficients for Y (coded units)

Term	Effect	Coef
Constant		18.000
Gas Flow	9.000	4.500
Intensity	-1.000	-0.500
Speed	1.500	0.750
Gas Flow*Intensity	-6.500	-3.250
Gas Flow*Speed	-0.500	-0.250
Intensity*Speed	1.000	0.500
Gas Flow*Intensity*Speed	0.500	0.250

Values of the main effects and the interactions together with the estimated model coefficients are shown. Note that each coefficient is a half of the corresponding effect.

. . .

Estimated Coefficients for Y using data in uncoded units

Term	Coef
Constant	-893.750
Gas Flow	102.625
Intensity	3.75000
Speed	186.250
Gas Flow*Intensity	-0.425000
Gas Flow*Speed	-30.0000
Intensity*Speed	-0.750000
Gas Flow*Intensity*Speed	0.125000

Model coefficients, in case uncoded units are used.

Normal Plot of the Effects
(response is Y, Alpha = 0,05)

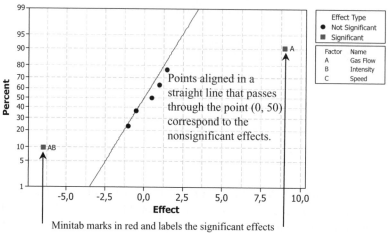

Points aligned in a straight line that passes through the point (0, 50) correspond to the nonsignificant effects.

Minitab marks in red and labels the significant effects

Half Normal Plot of the Effects
(response is Y, Alpha = 0,05)

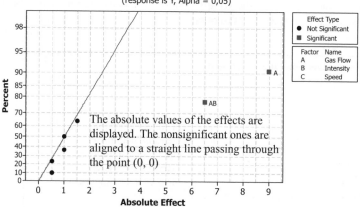

The absolute values of the effects are displayed. The nonsignificant ones are aligned to a straight line passing through the point (0, 0)

In some cases, the line drawn by Minitab to indicate which effects are significant may be inadequate; it tends to put it too much to the right and, thus to point as nonsignificant effects that might be significant. Therefore, to decide which effects are significant it is advised to complement Minitab indication with your own criterion.

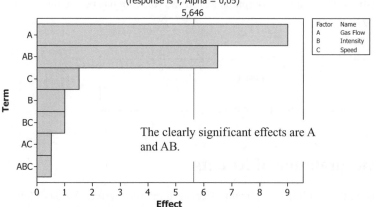

Minitab marks the border from which the effects are considered significant using an estimation of the standard deviation of the effects (Lenth method).

In many cases, a dotplot is enough to discriminate among significant and nonsignificant effects. To achieve that, proceed as follows:

1. Save the effects in a column using **Stat > DOE > Factorial > Analyze Factorial Design** and press **Storage.**

2. Construct the dotplot using **Graph > Dotplot.**

That is, choose **Stat > DOE > Factorial > Analyze Factorial Design > Storage** to save the effects:

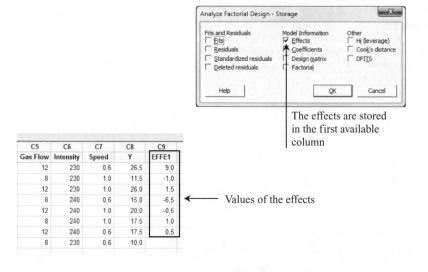

To build the dotplot, choose **Graph** > **Dotplot: Simple**, and place column C9 in **Graph variables.** The graphical output is as follows (the plot has been edited for the sake of clarity):

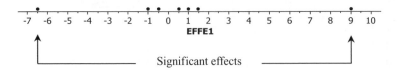

27.2   Interpretation of Results

In this example, the resulting significant effects are the main effect A (Gas Flow = 9) and the interaction AB (Gas Flow × Intensity = −6,5). Factor C (speed of the production chain) is inert.

Since factors A and B interact, it is necessary to draw an interaction plot to interpret the results. To do so, choose **Stat** > **DOE** > **Factorial** > **Factorial Plots**.

Use first the option **Interaction Plot**.

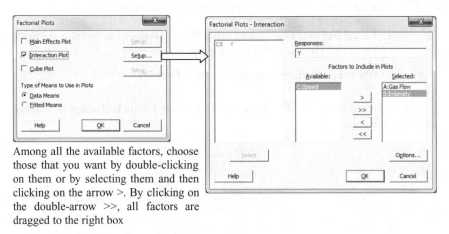

Among all the available factors, choose those that you want by double-clicking on them or by selecting them and then clicking on the arrow >. By clicking on the double-arrow >>, all factors are dragged to the right box

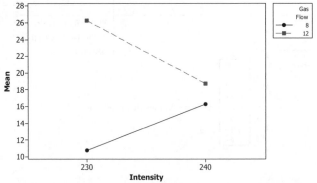

Use now the option **Cube Plot**. The displayed dialog box looks similar to the previous one.

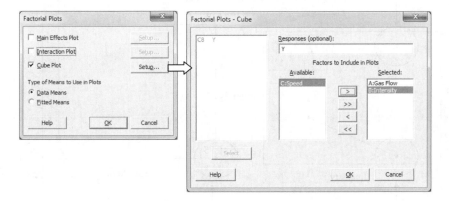

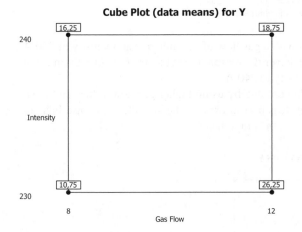

Cube Plot (data means) for Y

 Minitab draws very small numbers in the plot when using **Cube Plot**. If you want to use this plot in reports or presentations, it is recommended that you edit it by increasing the font size.

The two interaction plots are equivalent; from one you can build the other (the following plots have been edited to increase the font size of the numbers):

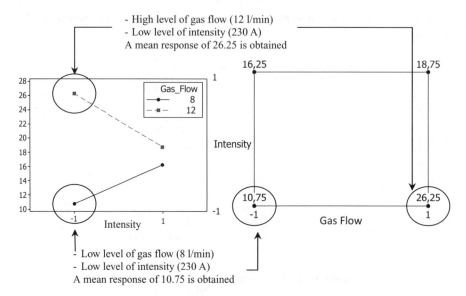

- High level of gas flow (12 l/min)
- Low level of intensity (230 A)
A mean response of 26.25 is obtained

- Low level of gas flow (8 l/min)
- Low level of intensity (230 A)
A mean response of 10.75 is obtained

The best result is obtained with a gas flow of 12 l/min and an intensity of 230 A. Note that if you ever need, whatever the reason, to set the gas flow to 8 l/min it will be better to increment the intensity to 240 A.

The experimentation could continue by using higher gas flow values and lower intensity values (if technically feasible) to explore the so-called response increase zone, as indicated by the arrow in the graph below.

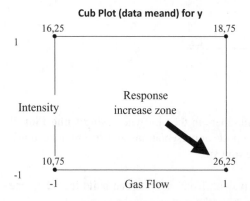

Cub Plot (data meand) for y

27.3   A Recap with a Fractional Factorial Design

Example 27.2: Consider again the second example of the previous chapter where a 2^{5-2} factorial design was carried out to improve the label's strength of adhesion of cava bottles. The five studied factors are: type of glue, temperature of glue, quantity of glue, drying temperature and brush pressure.

The Minitab worksheet window including the response variable looks as follows:

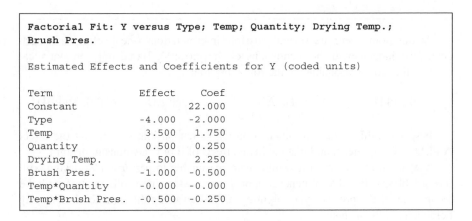

	C1	C2	C3	C4	C5-T	C6	C7	C8	C9	C10
	StdOrder	RunOrder	CenterPt	Blocks	Type	Temp	Quantity	Drying Temp.	Brush Pres.	Y
1	7	1	1	1	X	40	3	80	1,0	24,5
2	2	2	1	1	Y	30	2	80	1,0	16,0
3	8	3	1	1	Y	40	3	90	1,5	23,5
4	4	4	1	1	Y	40	2	90	1,0	24,5
5	3	5	1	1	X	40	2	80	1,5	22,5
6	1	6	1	1	X	30	2	90	1,5	24,0
7	5	7	1	1	X	30	3	90	1,0	25,0
8	6	8	1	1	Y	30	3	80	1,5	16,0

Choose **Stat > DOE > Factorial > Analyze Factorial Design**, using the column C10 as response and choosing the **Graphs** option to display the effects in a normal probability plot, to obtain:

```
Factorial Fit: Y versus Type; Temp; Quantity; Drying Temp.;
Brush Pres.

Estimated Effects and Coefficients for Y (coded units)

Term                Effect      Coef
Constant                      22.000
Type               -4.000     -2.000
Temp                3.500      1.750
Quantity            0.500      0.250
Drying Temp.        4.500      2.250
Brush Pres.        -1.000     -0.500
Temp*Quantity      -0.000     -0.000
Temp*Brush Pres.   -0.500     -0.250
```

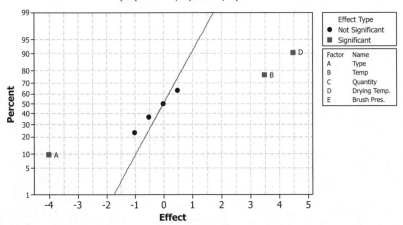

Normal Plot of the Effects
(response is Y, Alpha = 0,05)

The resulting significant effects are A+BD+CE, B+AD and D+AB.

 Although Minitab only labels the significant main effects on the plot, remember that a 2^{5-2} factorial design has resolution III, and thus the main effects are confounded with order two interactions.

The confounding pattern was previously obtained when the design matrix was created by choosing **Stat > DOE > Factorial > Create Factorial Design**, and appears again when analyzing the response.

```
I + ABD + ACE + BCDE

A + BD + CE + ABCDE
B + AD + CDE + ABCE
C + AE + BDE + ABCD
D + AB + BCE + ACDE
E + AC + BCD + ABDE
BC + DE + ABE + ACD
BE + CD + ABC + ADE
```

At this point there are several possible interpretations (the likelihood of each should be judged with the technicians of the process). Based only on statistical criteria, the more reasonable are the following four:

A, B and D   A, B, and AB   A, D and AD   B, D and BD

It seems unlikely that the interaction CE is responsible for making the effect A+BD+CE significant and that the main effects of E or C are nonsignificant.

In view of the situation it would seem reasonable to perform a 2^3 design with the variables A, B and D. Whenever appropriate, and again in collaboration with the technicians of the process, you might even think of using different levels that may help increase the response.

28

Response Surface Methodology

28.1 Matrix Design Creation and Data Collection

 Example 28.1: In a chemical reaction two components (Co1 and Co2) are mixed at a certain temperature (T) and speed (rpm). Among the reaction products that are obtained, there is one that is unwanted (Prod); the aim is to reduce its production to a minimum quantity.

The lasting time of the reaction is already known and remains constant, but it is decided to carry out a central composite design to see how the factors T, Co1, Co2 and rpm affect the response variable Prod.

Central composite designs are widely used because they allow us to plan the experimentation in two stages:

- A two level factorial design with centre points is carried out, checking for the need of quadratic terms.

- If a quadratic term is needed, 'star' points are added to allow the estimation of the pure quadratic terms.

Industrial Statistics with Minitab, First Edition. Pere Grima Cintas, Lluís Marco-Almagro and Xavier Tort-Martorell Llabrés.
© 2012 John Wiley & Sons, Ltd. Published 2012 by John Wiley & Sons, Ltd.

Stat > DOE > Response Surface > Create Response Surface Design

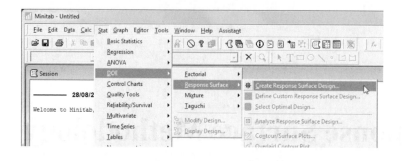

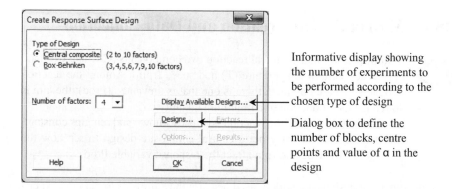

Informative display showing the number of experiments to be performed according to the chosen type of design

Dialog box to define the number of blocks, centre points and value of α in the design

28.1.1   Designs

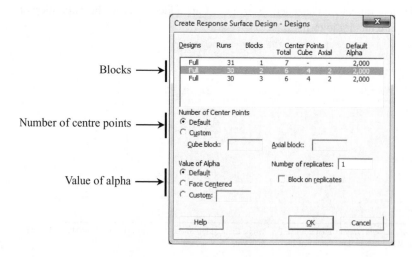

Blocks ———▶

Number of centre points ———▶

Value of alpha ———▶

Blocks: Choose the number of blocks. Usually, each block corresponds to sequentially performed experiments. For example, if experiments are carried out in two different days it is reasonable to have two blocks.

Number of Centre Points (in the cube and in the star): You can use the default option (recommended option) or choose the number you want.

Value of Alpha (distance of the star points to the centre of the design): You can use the default option (recommended option), locate the 'star' points at the centre of each face of the cube (**Face Centred**) or enter a specific value for alpha.

Once the design is chosen, proceed to name the factors, specify their corresponding low and high levels (**Factors** button), and choose if you want to randomize the order of experimentation (**Options** button).

28.1.2   Factors

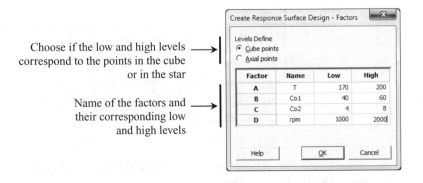

The design matrix in standard order (although the runs were done in randomized order), including the obtained response in each run, is the following:

C1	C2	C3	C4	C5	C6	C7	C8	C9	
StdOrder	RunOrder	PtType	Blocks	T	Co1	Co2	rpm	Prod	
1	23	1	1	170	40	4	1000	7.32	
2	12	1	1	200	40	4	1000	6.94	
3	11	1	1	170	60	4	1000	8.51	
4	17	1	1	200	60	4	1000	8.20	
5	21	1	1	170	40	8	1000	7.49	
6	30	1	1	200	40	8	1000	9.41	
7	29	1	1	170	60	8	1000	8.77	Experiments at
8	18	1	1	200	60	8	1000	9.54	the vertices of
9	13	1	1	170	40	4	2000	7.38	the cube
10	19	1	1	200	40	4	2000	8.97	
11	24	1	1	170	60	4	2000	9.29	
12	25	1	1	200	60	4	2000	8.06	
13	15	1	1	170	40	8	2000	8.68	
14	22	1	1	200	40	8	2000	6.94	
15	20	1	1	170	60	8	2000	10.50	
16	26	1	1	200	60	8	2000	10.94	
17	16	0	1	185	50	6	1500	9.82	Centre
18	28	0	1	185	50	6	1500	9.25	points of
19	14	0	1	185	50	6	1500	9.54	the cube
20	27	0	1	185	50	6	1500	9.94	
21	7	-1	2	155	50	6	1500	3.45	
22	9	-1	2	215	50	6	1500	3.44	
23	2	-1	2	185	30	6	1500	8.69	Experiments
24	4	-1	2	185	70	6	1500	12.11	at the star
25	5	-1	2	185	50	2	1500	9.02	
26	6	-1	2	185	50	10	1500	10.94	
27	10	-1	2	185	50	6	500	11.26	
28	1	-1	2	185	50	6	2500	9.26	Centre points
29	8	0	2	185	50	6	1500	9.14	of the star
30	3	0	2	185	50	6	1500	9.73	

If the design matrix is in random order, but you need it in standard order, go to **Stat > DOE > Display Design**.

If the design matrix is previously entered in the worksheet without using the option **Create Response Surface Design**, Minitab must be able to recognize such design matrix prior the statistical analysis. To achieve this, use the option **Stat > DOE > Response Surface > Define Custom Response Surface Design**. A similar situation appeared when dealing with factorial designs in Chapter 26.

28.2   Analysis of the Results

Stat > DOE > Response Surface > Analyze Response Surface Design

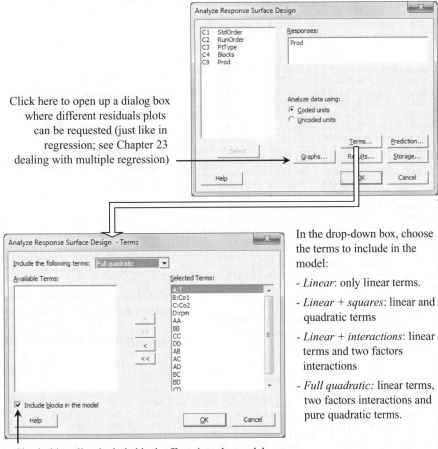

Click here to open up a dialog box where different residuals plots can be requested (just like in regression; see Chapter 23 dealing with multiple regression) →

In the drop-down box, choose the terms to include in the model:

- *Linear*: only linear terms.

- *Linear + squares*: linear and quadratic terms

- *Linear + interactions*: linear terms and two factors interactions

- *Full quadratic:* linear terms, two factors interactions and pure quadratic terms.

Check this cell to include block effects into the model

Selecting one of the options of the dropdown box, all those terms that correspond to the chosen option are placed in the list of **Selected Terms**. The central buttons with arrows can be used to move specific terms from the list of **Available Terms** to the list of **Selected Terms** for the model, or vice versa.

Start out with the full model (**Full quadratic**). The Minitab output has been divided in parts to ease the interpretation of it.

```
Response Surface Regression: Prod versus Block; T; Co1; Co2; rpm

The analysis was done using coded units.
Estimated Regression Coefficients for Prod

Term          Coef    SE Coef       T        P
Constant    9.55825    0.3673    26.025    0.000
Block       0.03525    0.1721     0.205    0.841
T           0.04333    0.1814     0.239    0.815
Co1         0.73000    0.1814     4.025    0.001
Co2         0.47667    0.1814     2.628    0.020
rpm         0.02417    0.1814     0.133    0.896
T*T        -1.52500    0.1697    -8.988    0.000
Co1*Co1     0.21375    0.1697     1.260    0.228
Co2*Co2     0.10875    0.1697     0.641    0.532
rpm*rpm     0.17875    0.1697     1.054    0.310
T*Co1      -0.10750    0.2221    -0.484    0.636
T*Co2       0.10750    0.2221     0.484    0.636
T*rpm      -0.18375    0.2221    -0.827    0.422
Co1*Co2     0.23625    0.2221     1.063    0.306
Co1*rpm     0.18500    0.2221     0.833    0.419
Co2*rpm    -0.05500    0.2221    -0.248    0.808
```

Values of the coefficients (just like the output of a regression analysis). Those coefficients with a high p-value can be removed from the model, one by one, as they will be non-significant.

```
S = 0.888594    PRESS = 71.3238
R-Sq = 89.54%   R-Sq(pred) = 32.53%   R-Sq(adj) = 78.34%
```

Hypothesis tests for sets of coefficients (linear terms, quadratic terms and interactions). If any of the p-values is large, none of the terms of the group is significant

The lack-of-fit test allows you to assess if the model provides a good fit. A small p-value in this test means that there is a lack of fit of the model

```
Analysis of Variance for Prod

Source           DF    Seq SS    Adj SS    Adj MS        F       P
Blocks            1     0.033    0.0331    0.0331     0.04   0.841
Regression       14    94.630   94.6299    6.7593     8.56   0.000
  Linear          4    18.302   18.3017    4.5754     5.79   0.006
  Square          4    73.929   73.9291   18.4823    23.41   0.000
  Interaction     6     2.399    2.3991    0.3998     0.51   0.794
Residual Error   14    11.054   11.0544    0.7896
  Lack-of-Fit    10    10.596   10.5959    1.0596     9.24   0.023
  Pure Error      4     0.459    0.4585    0.1146
Total            29   105.717

...
```

In the previous analysis of variance table a very large p-value (0.794) is reported for the interactions hypothesis test. This means that the null hypothesis (interactions are null) cannot be rejected and we proceed to remove them from the model.

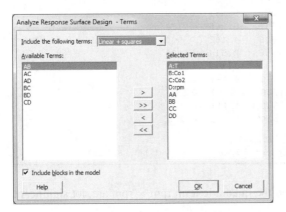

Linear and pure quadratic terms are selected

```
The analysis was done using coded units.
Estimated Regression Coefficients for Prod

Term          Coef   SE Coef        T       P
Constant    9.55825   0.3390   28.196   0.000
Block       0.03525   0.1588    0.222   0.827
T           0.04333   0.1674    0.259   0.798
Co1         0.73000   0.1674    4.360   0.000
Co2         0.47667   0.1674    2.847   0.010
rpm         0.02417   0.1674    0.144   0.887
T*T        -1.52500   0.1566   -9.738   0.000
Co1*Co1     0.21375   0.1566    1.365   0.187
Co2*Co2     0.10875   0.1566    0.694   0.495
rpm*rpm     0.17875   0.1566    1.141   0.267
```

The block effect is non-significant and is removed from the model

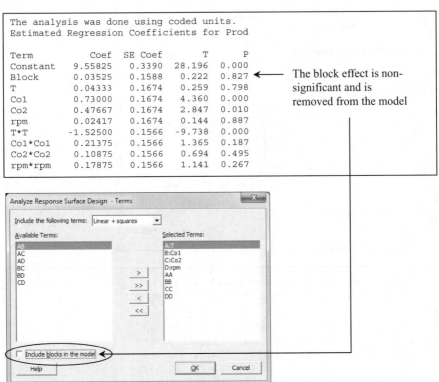

```
The analysis was done using coded units.
Estimated Regression Coefficients for Prod

Term          Coef   SE Coef        T       P
Constant    9.57000   0.3272   29.251   0.000
T           0.04333   0.1636    0.265   0.794
Co1         0.73000   0.1636    4.463   0.000
Co2         0.47667   0.1636    2.914   0.008
rpm         0.02417   0.1636    0.148   0.884
T*T        -1.52500   0.1530   -9.966   0.000
Co1*Co1     0.21375   0.1530    1.397   0.177
Co2*Co2     0.10875   0.1530    0.711   0.485
rpm*rpm     0.17875   0.1530    1.168   0.256

S = 0.801385    PRESS = 31.2041
R-Sq = 87.24%   R-Sq(pred) = 70.48%   R-Sq(adj) = 82.38%
```

Both the linear term rpm and the quadratic term rpm × rpm are not significant
and are removed from the model:

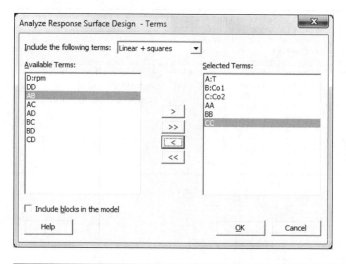

```
The analysis was done using coded units.
Estimated Regression Coefficients for Prod

Term          Coef   SE Coef        T       P
Constant    9.77429   0.2728   35.831   0.000
T           0.04333   0.1614    0.269   0.791
Co1         0.73000   0.1614    4.523   0.000
Co2         0.47667   0.1614    2.954   0.007
T*T        -1.55054   0.1494  -10.377   0.000
Co1*Co1     0.18821   0.1494    1.260   0.220
Co2*Co2     0.08321   0.1494    0.557   0.583

S = 0.790624    PRESS = 21.0578
R-Sq = 86.40%   R-Sq(pred) = 80.08%   R-Sq(adj) = 82.85%
```

Finally, the pure quadratic terms Co1 × Co1 and Co2 × Co2 are also removed from the model since they are not significant:

```
The analysis was done using coded units.
Estimated Regression Coefficients for Prod

Term        Coef   SE Coef       T       P
Constant  10.0156   0.1854   54.017   0.000
T          0.0433   0.1606    0.270   0.789
Co1        0.7300   0.1606    4.546   0.000
Co2        0.4767   0.1606    2.969   0.007
T*T       -1.5807   0.1466  -10.784   0.000

S = 0.786644    PRESS = 21.3908
R-Sq = 85.37%   R-Sq(pred) = 79.77%   R-Sq(adj) = 83.03%
```

The final model (in coded units) is thus:

$$\text{Prod} = 10.016 + 0.043\ \text{T} + 0.730\ \text{Co1} + 0.477\ \text{Co2} - 1.581\ \text{T}^2$$

 Although the p-value of the coefficient T is large, this coefficient is included in the model because T^2 must also appear. In fact, Minitab does not allow to remove T from the model if T^2 on it.

The plot of residuals versus predicted values (... **Analyze Response Surface Designs > Graphs > Residual Plots: Residuals versus fits**) shows no particular pattern in the final model. You can look at Chapter 23 on regression analysis for more information on residual analysis.

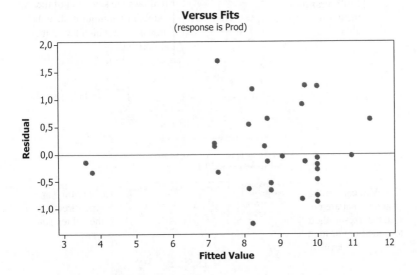

Versus Fits
(response is Prod)

28.3   Contour Plots and Response Surface Plots

When drawing the contour plots or response surfaces on the model chosen, you can only use 2 factors (one for each of the axes X and Y). The remaining factors should be set to one of their levels. A couple of suggestions are:

- If a factor is involved in a model in a quadratic form, it should be included as one of the axes.

- If a factor takes only a few values, a good idea is to produce a graph with other two factors for each of this levels.

Stat > DOE > Response Surface > Contour/Surface Plots
In our example, the variables T and Co1 are used for the axes, and the variable Co2 is set to its low, medium and high values. Since the variable rpm is not involved in the model, we should not worry about it.

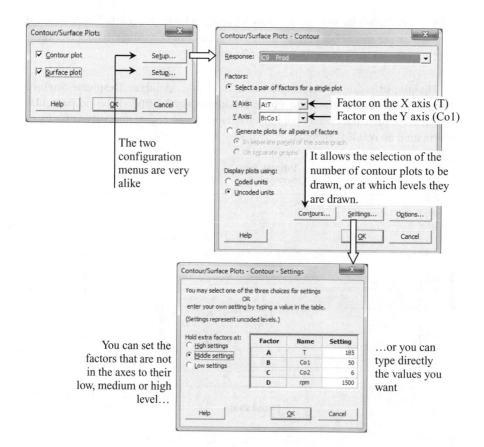

The following graphs show the contour plots and the response surfaces while maintaining the factor Co2 at its low, middle, and high level.

 Minitab draws contour plots and response surfaces of the latest model produced using **Analyze Response Surface Design**. It is not possible to memorize models and then draw them.

Component 2 (Co2) set to low level

Component 2 (Co2) set to middle level

Component 2 (Co2) set to high level

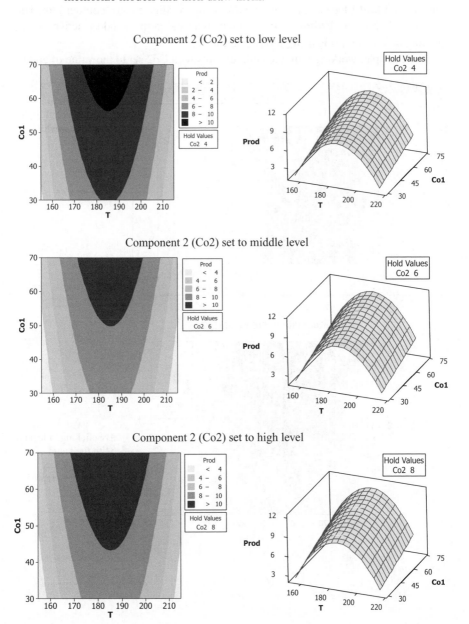

The surface looks like a tile that rises in the direction of Co1. The response variable also increases with the value of Co2 (if we had drawn T and Co2, we would also have bent tiles rising in the direction of Co2).

28.3.1   Score Points in the Contour Plots

The option **Plant Flag** can be used to place point coordinates on a contour plot. This is achieved choosing **Editor** > **Plant Flag** (with the graph window active), or by clicking on the **Graph Editing** toolbar button ⬚.

For example, working with the contour plots with Co2 set to the value of eight:

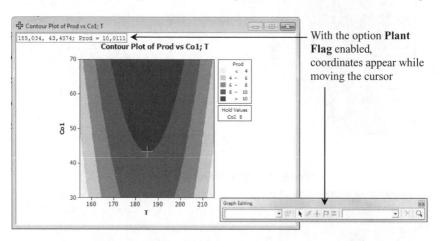

With the option **Plant Flag** enabled, coordinates appear while moving the cursor

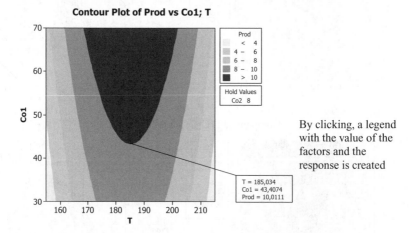

By clicking, a legend with the value of the factors and the response is created

29

Reliability

29.1 File

 A company manufactures injection pumps for diesel engines. The file INJECTION.MTW contains data on the lifetime (in hours) of 40 injection pumps.

A first approximation to the analysis of these data is obtained by drawing a histogram:

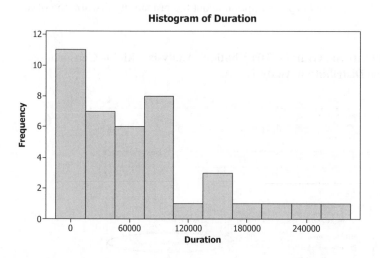

The data do not follow a normal distribution, which is somehow expected, as these data refer to lifetimes of equipment. Probably an exponential or Weibull distribution can be fit to these data.

Industrial Statistics with Minitab, First Edition. Pere Grima Cintas, Lluís Marco-Almagro and Xavier Tort-Martorell Llabrés.
© 2012 John Wiley & Sons, Ltd. Published 2012 by John Wiley & Sons, Ltd.

Focus now in the case of complete data (the failure times of all elements are known) or in the case of right censored data (it is known that some elements have lasted more than a certain time, because when the experiment ended they were still functioning).

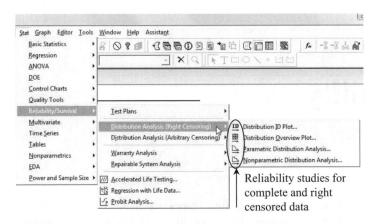

Reliability studies for complete and right censored data

29.2   Nonparametric Analysis

In the nonparametric analysis, the reliability and risk functions are empirical, calculated from the data, without any assumption about the probability distribution of the data.

Stat > Reliability/Survival > Distribution Analysis (Right Censoring) > Nonparametric Distribution Analysis

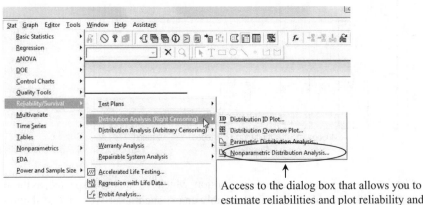

Access to the dialog box that allows you to estimate reliabilities and plot reliability and risk functions for complete or right censored data without assuming any theoretical model

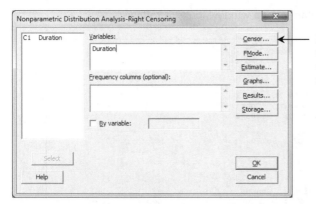

No changes needed if the
data are complete. Click
on **Censor...** in case of
right censored data

29.2.1 Graphs

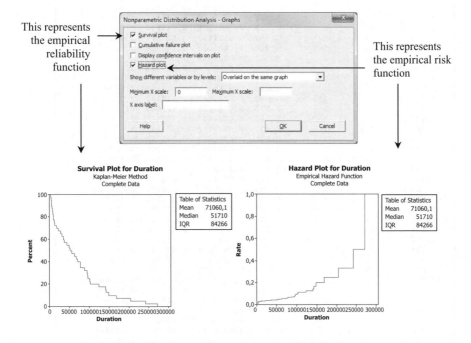

This represents
the empirical
reliability
function

This represents
the empirical risk
function

The output in the Minitab session window looks as follows:

```
...

Characteristics of Variable

            Standard    95,0% Normal CI
Mean(MTTF)    Error    Lower     Upper
   71060.1  10634.4  50216.9   91903.2

Median = 51710
IQR = 84266   Q1 = 12504   Q3 = 96770

Kaplan-Meier Estimates


Number  Number   Survival   Standard    95,0% Normal CI
 Time   at Risk  Failed  Probability    Error     Lower      Upper
  3607     40      1       0.975    0.0246855  0.926617  1.00000
  4100     39      1       0.950    0.0344601  0.882459  1.00000
  5734     38      1       0.925    0.0416458  0.843376  1.00000
  5768     37      1       0.900    0.0474342  0.807031  0.99297
  7025     36      1       0.875    0.0522913  0.772511  0.97749
  8089     35      1       0.850    0.0564579  0.739344  0.96066
  9411     34      1       0.825    0.0600781  0.707249  0.94275
 10640     33      1       0.800    0.0632456  0.676041  0.92396
 10681     32      1       0.775    0.0660256  0.645592  0.90441
 12504     31      1       0.750    0.0684653  0.615810  0.88419
 13030     30      1       0.725    0.0706001  0.586626  0.86337
 17656     29      1       0.700    0.0724569  0.557987  0.84201
 22339     28      1       0.675    0.0740566  0.529852  0.82015
 28698     27      1       0.650    0.0754155  0.502188  0.79781
 31749     26      1       0.625    0.0765466  0.474972  0.77503
 34585     25      1       0.600    0.0774597  0.448182  0.75182
 36863     24      1       0.575    0.0781625  0.421804  0.72820
 43403     23      1       0.550    0.0786607  0.395828  0.70417
 49389     22      1       0.525    0.0789581  0.370245  0.67975
 51710     21      1       0.500    0.0790569  0.345051  0.65495
 56084     20      1       0.475    0.0789581  0.320245  0.62975
 63311     19      1       0.450    0.0786607  0.295828  0.60417
 68135     18      1       0.425    0.0781625  0.271804  0.57820
 71329     17      1       0.400    0.0774597  0.248182  0.55182
 77223     16      1       0.375    0.0765466  0.224972  0.52503
 77629     15      1       0.350    0.0754155  0.202188  0.49781
 87564     14      1       0.325    0.0740566  0.179852  0.47015
 94596     13      1       0.300    0.0724569  0.157987  0.44201
 96104     12      1       0.275    0.0706001  0.136626  0.41337
 96770     11      1       0.250    0.0684653  0.115810  0.38419
101214     10      1       0.225    0.0660256  0.095592  0.35441
102993      9      1       0.200    0.0632456  0.076041  0.32396
123815      8      1       0.175    0.0600781  0.057249  0.29275
140341      7      1       0.150    0.0564579  0.039344  0.26066
142312      6      1       0.125    0.0522913  0.022511  0.22749
148521      5      1       0.100    0.0474342  0.007031  0.19297
168021      4      1       0.075    0.0416458  0.000000  0.15662
204471      3      1       0.050    0.0344601  0.000000  0.11754
242796      2      1       0.025    0.0246855  0.000000  0.07338
272193      1      1       0.000    0.0000000  0.000000  0.00000
```

| Moment of failure | Number of injection pumps working until a new unit fails | Number of units failing at each moment | Empirical reliability value (number of units still working among the initially available ones) |

Without specifying any model, that is, using a nonparametric analysis, what is the reliability after 40 000 functioning hours? In the previous table, take a look at the row corresponding to 36 863 hours (displayed in bold). The next failure was produced at 43 403 hours; thus, the units still in function after 40 000 hours are the same that were working after 36 863 hours. Therefore, the reliability value at 40 000 hours is equal to 0.575.

29.3 Identification of the Best Model for the Data

Stat > Reliability/Survival > Distribution Analysis (Right Censoring) > Distribution ID Plot

Column containing the failure times —→

Either produce the probabilistic plots with all available distributions or choose the most adequate ones for the data —→

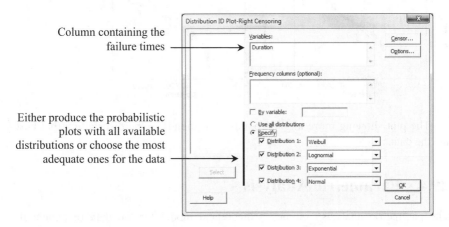

The first block on the Minitab output session window displays the goodness of fit statistics that will allow you to find the best model for the data:

Distribution ID Plot: Duration

Goodness-of-Fit

Distribution	Anderson-Darling (adj)	Correlation Coefficient
Weibull	0.776	0.977
Lognormal	1.072	0.977
Exponential	0.654	*
Normal	1.860	0.931

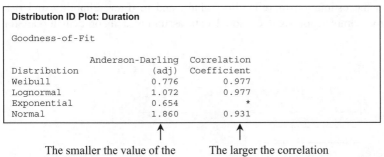

The smaller the value of the Anderson-Darling statistic, the better the fit

The larger the correlation coefficient, the better the fit

Probability Plot for Duration
LSXY Estimates-Complete Data

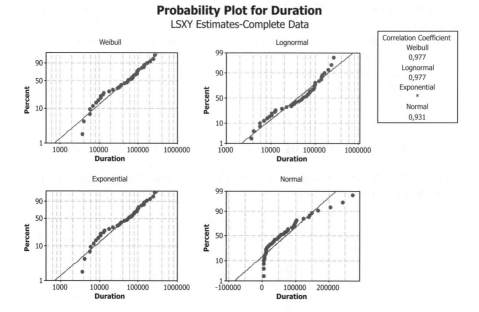

The plot showing more aligned points corresponds to the distribution that best fits the data.

29.4   Parametric Analysis

The parametric analysis assumes a theoretical model for the data (exponential, Weibull, lognormal, etc.). The identification of the distribution that best fits the data is achieved through **Distribution ID Plot**, as previously seen.

Stat > Reliability/Survival > Distribution Analysis (Right Censoring) > Parametric Distribution Analysis
This allows the estimation of reliability values, and to draw reliability and risk functions for complete or right censored data assuming a theoretical model for the data.

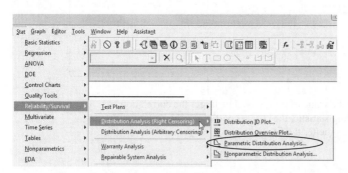

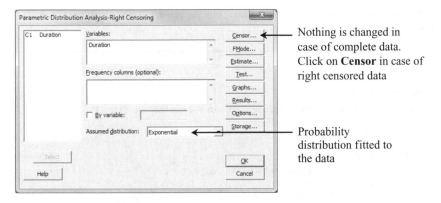

Nothing is changed in case of complete data. Click on **Censor** in case of right censored data

Probability distribution fitted to the data

29.4.1   Graphs

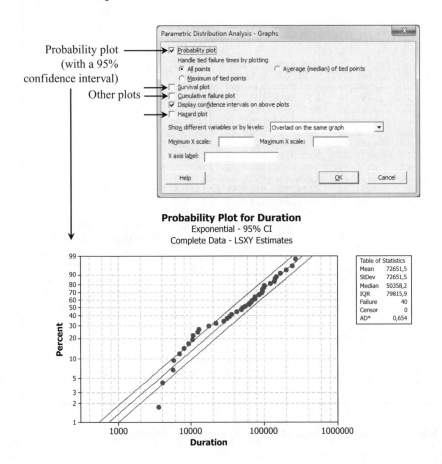

Probability plot (with a 95% confidence interval)

Other plots

The session window after carrying out a **parametric analysis** looks as follows:

```
Variable: Duration

Censoring Information  Count
Uncensored value         40

Estimation Method: Least Squares (failure time(X) on rank(Y))

Distribution:   Exponential

Parameter Estimates

                      Standard    95.0% Normal CI
Parameter  Estimate     Error    Lower     Upper
Mean        72651.5   11615.1   53107.9   99387.2

Log-Likelihood = -486.861

Goodness-of-Fit
Anderson-Darling (adjusted) = 0.654

...


Table of Percentiles

                      Standard    95,0% Normal CI
Percent  Percentile     Error    Lower     Upper
     1     730.172    116.736   533.752   998.875
     2    1467.76     234.657  1072.92   2007.89
     3    2212.91     353.788  1617.63   3027.26
     4    2965.78     474.153  2167.97   4057.18
     5    3726.54     595.779  2724.08   5097.90
     6    4495.34     718.692  3286.07   6149.62
     7    5272.37     842.919  3854.08   7212.60
     8    6057.80     968.489  4428.22   8287.06
     9    6851.82    1095.43   5008.64   9373.27
    10    7654.60    1223.78   5595.48  10471.5
    20   16211.7     2591.84  11850.7   22177.6
    30   25913.0     4142.83  18942.3   35448.9
    40   37112.3     5933.31  27128.9   50769.5
    50   50358.2     8051.00  36811.6   68890.0
    60   66569.9    10642.8   48662.3   91067.6
    70   87470.5    13984.3   63940.5  119659
    80  116928      18693.8   85473.9  159958
    90  167286      26744.9  122286    228847
    91  174941      27968.6  127881    239319
    92  183498      29336.7  134136    251025
    93  193199      30887.7  141228    264296
    94  204399      32678.2  149414    279617
    95  217645      34795.9  159097    297737
    96  233856      37387.7  170948    319915
    97  254757      40729.2  186226    348507
    98  284215      45438.7  207759    388805
    99  334573      53489.7  244571    457695
```

Values calculated from the theoretical model. After 7655 hours 10% of the injection pumps do not function. Thus, the reliability value at 7655 hours is equal to 90%.

What is the reliability value after 40 000 hours of functioning?

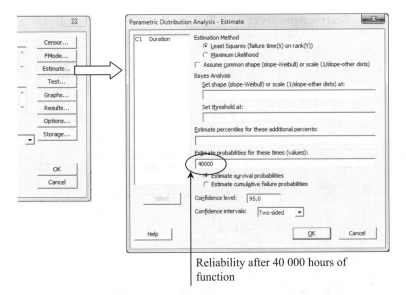

Reliability after 40 000 hours of function

The Minitab output is:

```
Table of Survival Probabilities

                      95,0% Normal CI
  Time   Probability    Lower      Upper
 40000     0.576619   0.470865   0.668669
```

Thus, the reliability value at 40 000 hours is equal to 0.5766. The reliability value obtained (at 40 000 hours) from a parametric (assuming an exponential model) or nonparametric (without assuming any model) analysis can be compared.

- Nonparametric: Reliability value at 40 000 hours = 0.5750

- Parametric: Reliability value at 40 000 hours = 0.5766

29.5   General Graphical Display of Reliability Data

A quick way to produce the reliability and risk functions, from either a parametric or nonparametric analysis, is via: **Stat > Reliability/Survival > Distribution Analysis (Right Censoring) > Distribution Overview Plot**

In **Variables** place the column C1 (containing lifetime data). Enable the **Parametric analysis** option, with the Exponential distribution, to obtain:

Distribution Overview Plot for Duration
LSXY Estimates-Complete Data

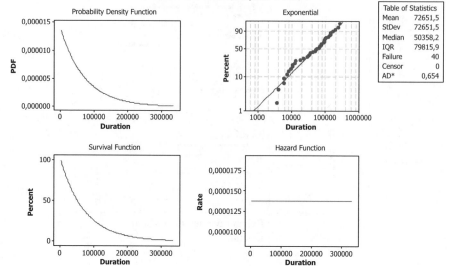

Table of Statistics	
Mean	72651,5
StDev	72651,5
Median	50358,2
IQR	79815,9
Failure	40
Censor	0
AD*	0,654

With the nonparametric analysis, the plots are:

Distribution Overview Plot for Duration
Kaplan-Meier Estimates-Complete Data

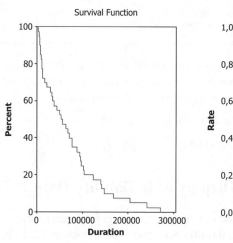

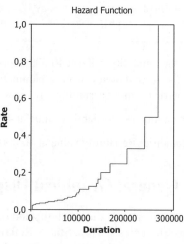

30

Part Six: Case Studies
Design of Experiments and Reliability

30.1 Cardigan

The front part of a cardigan consists of two halves, each of which is composed of a body and a strip. The strip is the part where the buttons or the buttonholes are (depending on the side) and it is woven in such a way that it is more consistent and stronger than the body.

Traditionally, body and strip are woven separately and then sewn together, but nowadays there are machines that weave simultaneously body and strip as a single part, the advantage being that operations during the manufacture of the garment are eliminated. The drawback is that since the strip has a different stitch and is woven under other parameters, it sometimes ends up being longer or shorter than the body, resulting in a defective part.

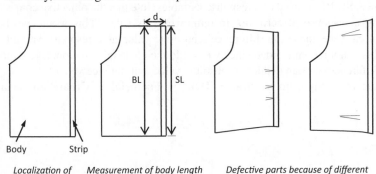

Localization of body and strip

Measurement of body length (BL) and strip length (SL).The distance (d) is fixed.

Defective parts because of different body and strip lengths

Industrial Statistics with Minitab, First Edition. Pere Grima Cintas, Lluís Marco-Almagro and Xavier Tort-Martorell Llabrés.
© 2012 John Wiley & Sons, Ltd. Published 2012 by John Wiley & Sons, Ltd.

To determine the weaving conditions for the strip that guarantee that the length of body and strip are the same, a 2^3 design is carried out with the following factors:

A: Type of strip: Interlock (−) and tubular (+)
B: Number of needles: 4 and 10
C: Graduation of the stitch (length of the mesh in the interior part): 9.0 and 10.8

The following results are obtained, in the standard order of the design matrix:

Weaving time of the part (min):	12:40	12:39	12:39	12:39	9:18	9:18	9:17	9:18
Body Length (BL, cm):	67.8	71.7	67.6	77.0	62.2	71.6	71.7	75.6
Strip Length (SL, cm):	70.4	70.2	70.0	75.9	64.9	73.8	74.4	78.0

The objective is to determine the weaving conditions such that the time and the difference between body and strip length are minimized. It is of particular interest to find out whether it is possible to weave in less than 10 minutes with a zero difference between both lengths.

Concerning the time needed to weave the part, it can be seen that the first four values as well as the last four ones are nearly identical. Hence, without making any calculations, we can state that the time depends only on factor C: going from level − to level +, that is from 9 to 10.8 units of stitch graduation, the weaving time decreases by approximately 1:21 min. Thus, in order to minimize the time, we are interested in the high level of stitch graduation.

In order to find out how the studied factors influence the difference of the lengths, the difference, SL-BL, is used as a response variable. Note that the objective consists in achieving a response of zero, not in minimizing its value. This would be the objective in case we chose the difference's absolute value as a response variable; however, this is not recommended since we would then lose information. (Note that given the difference, we can deduce its absolute value, but not vice versa.)

To create the design, go to: **Stat > DOE > Factorial > Create Factorial Design**

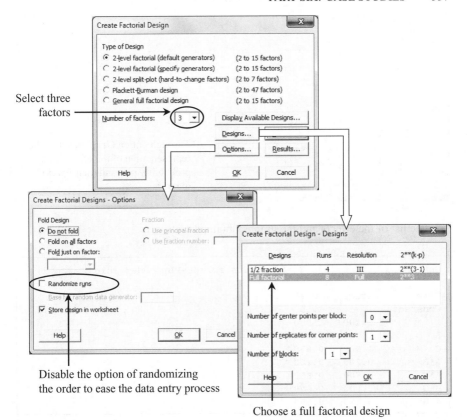

Select three
factors

Disable the option of randomizing
the order to ease the data entry process

Choose a full factorial design

Introduce both measures, body length (BL) and strip length (SL), to the worksheet and calculate their difference to obtain:

↓	C1	C2	C3	C4	C5	C6	C7	C8	C9	C10
	StdOrder	RunOrder	CenterPt	Blocks	A	B	C	BL	SL	SL-BL
1	1	1	1	1	-1	-1	-1	68,7	70,4	2,6
2	2	2	1	1	1	-1	-1	71,7	70,2	-1,5
3	3	3	1	1	-1	1	-1	67,6	70,0	2,4
4	4	4	1	1	1	1	-1	77,0	75,9	-1,1
5	5	5	1	1	-1	-1	1	62,2	64,9	2,7
6	6	6	1	1	1	-1	1	71,6	73,8	2,2
7	7	7	1	1	-1	1	1	71,7	74,4	2,7
8	8	8	1	1	1	1	1	75,6	78,0	2,4

Analyze the results using: **Stat > DOE > Factorial > Analyze Factorial Design**

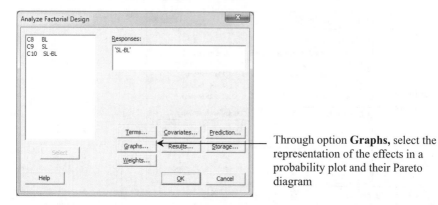

Through option **Graphs,** select the representation of the effects in a probability plot and their Pareto diagram

The list of the effects and the graphs obtained are:

```
Estimated Effects and Coefficients for SL-BL (coded units)

Term       Effect      Coef
Constant               1,550
A         -2.100     -1.050
B          0.100      0.050
C          1.900      0.950
A*B        0.200      0.100
A*C        1.700      0.850
B*C       -0.000     -0.000
A*B*C     -0.100     -0.050
```

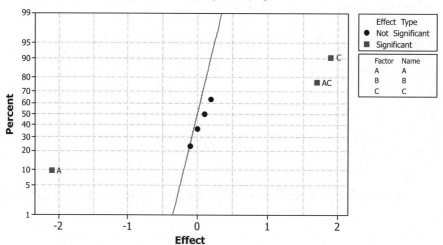

Pareto Chart of the Effects
(response is SL-BL, Alpha = 0,05)

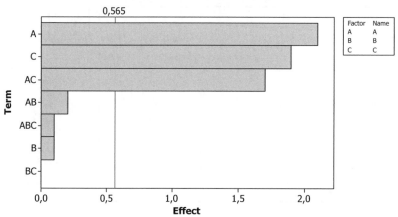

From both graphs it becomes clear that the significant effects are the main effects of A (type of strip) and C (stitch graduation) as well as their interaction AC.

To interpret the result, we directly take a look at the graph of the interaction AC:

Stat > DOE > Factorial > Factorial Plots

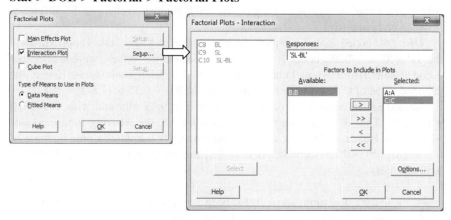

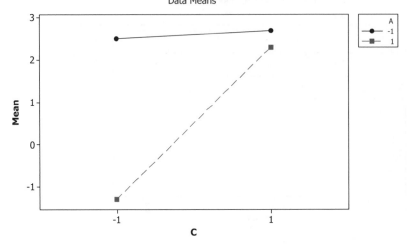

Interaction Plot for SL-BL
Data Means

The number of needles (B) is inert for both answers of interest (time of weaving and difference between body and strip). Concerning the two other factors, the interaction plot indicates that with strip type Interlock (A, level −) the difference remains almost constant and larger than 2 mm, regardless the value of C. However, in case of strip type tubular (A, level +), changing the stitch graduation changes the difference between body and strip.

Using the coded values, a zero-response is obtained when C takes approximately the value C = −0.25 (under A level +). To decode the values, we consider that 2 coded units (from −1 to 1) are equivalent to 1.8 stitch graduations. Hence, the decoded value of C that gives a zero difference is: $9.0 + 0.75 \times (1.8/2) = 9.7$.

Regarding the weaving time, with C = 9 it is 9:18 min and with C = 10.8 it is 12:39. That is, increasing the graduation by 1.8, the weaving time increases 81 seconds. Hence, an increase of the graduation by 0.7, implies a time increase of $0.7 \times (81/1.8) = 31.5$ seconds. At C = 9.7, the weaving time is $9:18 + 0:32 = 9:50$ minutes.

In the calculations above, it has been assumed that the response has a linear behaviour between the low and high levels.

30.2   Steering wheel – 1

A manufacturer of steering wheels for cars had problems with the hardness of its product (a critical characteristic of the steering wheels is that they must be sufficiently hard in order not to break, as well as sufficiently soft so that they break in case of an accident and do not harm the ribs of the driver). The manufacturing process consists of injecting polyurethane into a mould.

To find out what the breakage index depends on, a 2^3 experiment is carried out with variables P (Injection pressure), R (ratio of the two components of the polyurethane), and T (Injection temperature). After properly choosing the levels and given the large variability detected in the hardness, it was decided to replicate the experiment. The obtained results are:

P	R	T	Hardness1	Hardness2
−1	−1	−1	35	18
1	−1	−1	62	47
−1	1	−1	28	31
1	1	−1	55	56
−1	−1	1	49	26
1	−1	1	48	31
−1	1	1	34	39
1	1	1	45	44

The aim consists in analyzing how each of the factors affects the hardness.

Start by creating the factorial design.

Stat > DOE > Factorial > Create Factorial Design

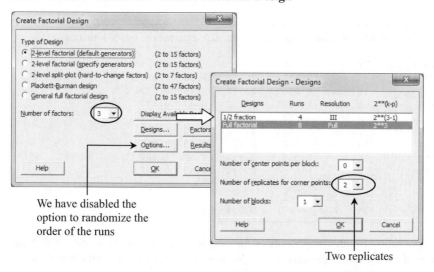

We have disabled the
option to randomize the
order of the runs

Two replicates

In replicated designs, the design matrix is repeated as many times as the number of replicates. First, the first replicates are entered, then, the second replicates, etc. Thus, all the response values are placed in the same column.

	C1	C2	C3	C4	C5	C6	C7	C8	
Worksheet 1 ***	StdOrder	RunOrder	CenterPt	Blocks	P	R	T	Hardness	
1	1	1	1	1	-1	-1	-1	35	First replicates
2	2	2	1	1	1	-1	-1	62	
3	3	3	1	1	-1	1	-1	28	
4	4	4	1	1	1	1	-1	55	
5	5	5	1	1	-1	-1	1	49	
6	6	6	1	1	1	-1	1	48	
7	7	7	1	1	-1	1	1	34	
8	8	8	1	1	1	1	1	45	
9	9	9	1	1	-1	-1	-1	18	Second replicates
10	10	10	1	1	1	-1	-1	47	
11	11	11	1	1	-1	1	-1	31	
12	12	12	1	1	1	1	-1	56	
13	13	13	1	1	-1	-1	1	26	
14	14	14	1	1	1	-1	1	31	
15	15	15	1	1	-1	1	1	39	
16	16	16	1	1	1	1	1	44	

Analyze the results using: **Stat > DOE > Factorial > Analyze Factorial Design**

In this case, it is not necessary to analyze the significance of the effects using graphs because Minitab, as there are replicates, carries out significance tests for each of the effects.

Setting a 5% significance level as a criterion, the significant effects are the main effect of P and the PT interaction.

```
Estimated Effects and Coefficients for Hardness (coded units)

Term         Effect     Coef   SE Coef       T      P
Constant              40,500     2,312   17,52  0,000
P            16.000    8.000     2.312    3.46  0.009
R             2.000    1.000     2.312    0.43  0.677
T            -2.000   -1.000     2.312   -0.43  0.677
P*R           1.000    0.500     2.312    0.22  0.834
P*T         -11.000   -5.500     2.312   -2.38  0.045
R*T          -0.000   -0.000     2.312   -0.00  1.000
P*R*T         2.000    1.000     2.312    0.43  0.677


S = 9.24662      PRESS = 2736
R-Sq = 69.52%    R-Sq(pred) = 0.00%   R-Sq(adj) = 42.85%
```

The ratio of both components of the polyurethane (factor R) is inert within the range of values used in the experiment. We now have a look at the interaction between pressure and the injection temperature (PT):

Stat > DOE > Factorial > Factorial Plots

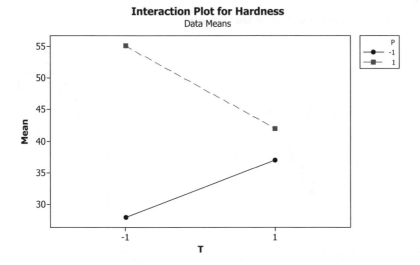

Interaction Plot for Hardness
Data Means

The maximum resistance is obtained with pressure level + and the temperature level –. If at any time it was necessary to work with pressure level –, the temperature should be set at level +.

30.3   Steering Wheel – 2

In a later study on the hardness of the steering wheels, it was discovered that the previous experiment was not properly randomized. Initially, the first 8 replicates were done, and then the other 8, with the aggravating circumstance that two weeks passed between both sets of experiments; it is known that the environmental conditions (temperature, humidity, . . .) affect the characteristics of the components of polyurethane.

Because of the circumstances under which the data were obtained, it was decided to re-analyze the experiment considering it a 2^4 design where a new factor W (level -1 for the 8 experiments of the first replicate, and level + for the 8 of the second one) represents the differences (environmental or of other type) occurred during the two weeks passed between the first and the second experiment.

It would be interesting to draw conclusions considering this new point of view.

Stat > DOE > Factorial > Create Factorial Design
We have 4 factors; choose the complete design without replicates, and disable the option to randomize the order of the runs. Since we do not have replicates now, we activate the options of drawing a probability plot and a Pareto diagram of the effects to analyze its significance.

Stat > DOE > Factorial > Analyze Factorial Design > Graphs > Effects Plots: Normal

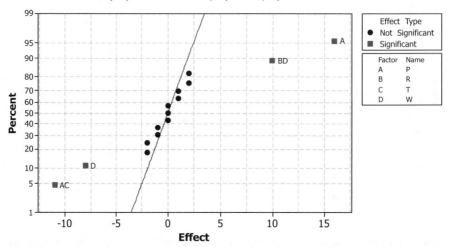

It is clear from both graphs that the significant effects are A (P), AC (PT) (up to that point, it is the same result as in the previous analysis), D (W), and BD (RW). Analyzing the PT interaction, the conclusions are the same as before. Analyzing the RW interaction, we obtain the following.

Stat > DOE > Factorial > Factorial Plots

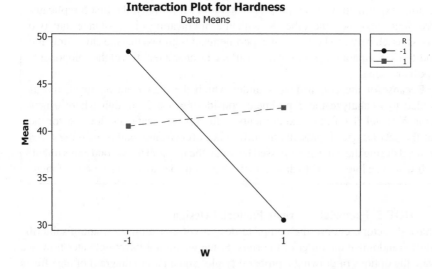

If R is at level −, the environmental factor clearly affects the hardness of the steering wheels; however, if it is at level +, this influence is much less. Previously, when the environmental conditions were ignored, R appeared as inert; hence, its value was indifferent. Now, in view of its interaction with the environmental conditions, it can be used to neutralize the influence of these and get a product with features that are independent of the conditions under which it has been produced.

30.4   Paper Helicopters

Following the idea presented in the paper by George Box, 'Teaching Engineers Experimental Design with a Paper Helicopter', *Quality Engineering* **4**(3) (1992), it is possible to construct a paper helicopter following the scheme in the figure below. Dropping it from a certain height, it first falls down in an irregular way, but when it opens the wings, it starts to rotate around itself falling down slowly.

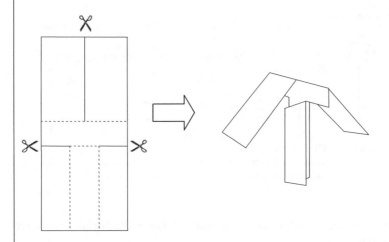

To improve the design with the aim to maximize the flying time, a 2^{8-4} factorial design is carried out with the following factors and levels:

	Factor	−	+
A	Type of paper	Normal	Strong
B	Body length	5 cm	6.5 cm
C	Body width	3 cm	4 cm
D	Wing length	5 cm	6.25 cm
E	Overweight (clip)	No	Yes
F	Body pasted	No	Yes
G	End of the wing bent upwards	No	Yes
H	Adhesive tape on the wing	No	Yes

To write the design matrix, the generators provided by Minitab were used, and the flying times (in seconds), in standard order, were:

2.3; 2.2; 2.6; 2.3; 3.1; 2.9; 2.5; 2.6; 1.9; 2.1; 1.7; 1.8; 2.1; 1.9; 2.3; 2.4

The objective is drawing conclusions from the experiments performed, and suggesting which should be the next experiments to be carried out.

To create the design go to: **Stat > DOE > Factorial > Create Factorial Design**

We have 8 factors and choose a 2^{8-4} design. Proceed to disable the option to randomize runs and, for the analysis, choose the graph of the effects in normal probability plot.

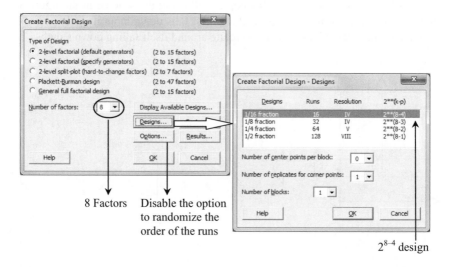

2^{8-4} design

Introduce the results and analyze the data: **Stat > DOE > Factorial > Analyze Factorial Design**

Factorial Fit: Y versus A; B; C; D; E; F; G; H

```
Estimated Effects and Coefficients for Y (coded units)


Term          Effect      Coef
Constant                2,2938
A            -0.0375   -0.0187
B            -0.0375   -0.0187
C             0.3625    0.1812
D            -0.5375   -0.2687
E             0.3125    0.1563
F            -0.0875   -0.0437
G             0.1125    0.0563
H             0.0125    0.0063
A*B           0.0375    0.0187
A*C          -0.0125   -0.0063
A*D           0.0875    0.0437
A*E          -0.0125   -0.0063
A*F          -0.0625   -0.0313
A*G          -0.0125   -0.0063
A*H           0.0875    0.0437


...
Alias Structure (up to order 3)
I
A + B*C*G + B*D*H + B*E*F + C*D*F + C*E*H + D*E*G + F*G*H
B + A*C*G + A*D*H + A*E*F + C*D*E + C*F*H + D*F*G + E*G*H
C + A*B*G + A*D*F + A*E*H + B*D*E + B*F*H + D*G*H + E*F*G
D + A*B*H + A*C*F + A*E*G + B*C*E + B*F*G + C*G*H + E*F*H
E + A*B*F + A*C*H + A*D*G + B*C*D + B*G*H + C*F*G + D*F*H
F + A*B*E + A*C*D + A*G*H + B*C*H + B*D*G + C*E*G + D*E*H
G + A*B*C + A*D*E + A*F*H + B*D*F + B*E*H + C*D*H + C*E*F
H + A*B*D + A*C*E + A*F*G + B*C*F + B*E*G + C*D*G + D*E*F
A*B + C*G + D*H + E*F
A*C + B*G + D*F + E*H
A*D + B*H + C*F + E*G
A*E + B*F + C*H + D*G
A*F + B*E + C*D + G*H
A*G + B*C + D*E + F*H
A*H + B*D + C*E + F*G
```

Representing the effects in a normal probability plot, the following plot is obtained:

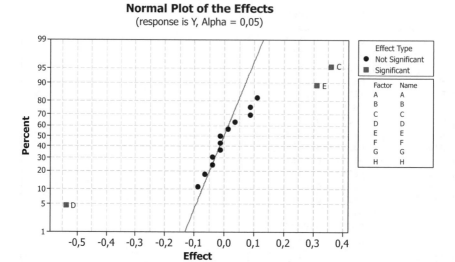

Normal Plot of the Effects
(response is Y, Alpha = 0,05)

The only effects that appear to be significant are the main effects of C (Body width), D (Wing length), and E (Overweight). When choosing the design we already saw that our 2^{8-4} design is of resolution IV. That means that the main effects are only confounded with 3-factors interactions. This confounding can be seen in detail in the alias structure obtained when analyzing the results.

Discarding the interactions of 3 or more factors, the conclusions are that increasing the body length and adding overweight improve the flying time, while increasing the wing length worsens it. The remaining factors under study are inert within the levels of the experiment.

For another experiment, it might be a good option to use a 2^3 design with the 3 factors that appeared to have a significant effect, moving the levels towards the direction that seems to be more promising. That is:

Factor		−	+
C	Body width	4 cm	5 cm
D	Wing length	4 cm	5 cm
E	Overweight (clip)	1 clip	2 clips

30.5   Microorganisms

In the laboratory of a company of wastewater treatment an experiment is carried out with the aim of detecting how the pH of the water and the temperature affect the microorganisms' growth rate. In particular, the aim is to find out which pH value and temperature cause the highest growth rate.

The experiments will be performed following the response surface methodology, according to the next steps:

1. A 2^2 design with centre points is carried out in order to try to fit a linear model. A good model fit would indicate we are far from the maximum.

2. Both factors are moved in the direction of the maximum growth (steepest ascent), to detect where the maximum is.

3. A new 2^2 design with centre points is carried out and a linear model is fit. In case of a good fit, we are still far from the maximum and, hence, the ascent direction will be corrected and we continue looking for the maximum.

4. In case the previous model does not fit well, the maximum has probably already been reached. We then add more experiments to be able to estimate the quadratic terms.

→ **Step 1**

We start with a 2^2 design with 3 center points, with the low and high levels for pH and temperature that are shown in the following table

	Level −	Level +
pH	3,0	4,0
Temperature	20	25

The design matrix with the obtained results:

pH	Temp [°C]	Rate [$\cdot 10^{-2}$ min^{-1}]
3.0	20.0	1.56
4.0	20.0	2.55
3.0	25.0	2.30
4.0	25.0	3.20
3.5	22.5	2.45
3.5	22.5	2.31
3.5	22.5	2.31

Since we already introduced the design matrix and the responses in the worksheet, we do not use **Create Response Surface Design**, but **Custom Response Surface Design.**

Stat > DOE > Response Surface > Define Custom Response Surface Design

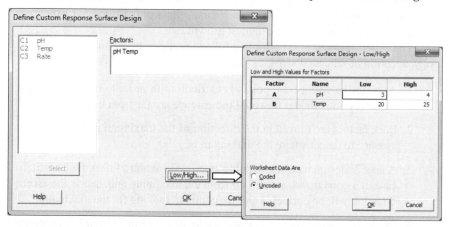

Stat > DOE > Response Surface > Analyze Response Surface Design

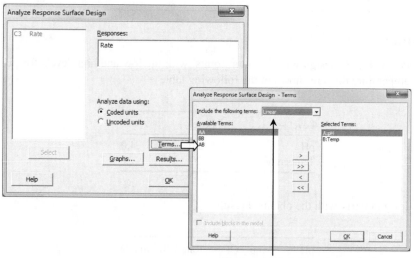

We only include linear terms in the model

Response Surface Regression: Rate versus pH; Temp

```
The analysis was done using coded units.

Estimated Regression Coefficients for Rate

Term          Coef   SE Coef      T       P
Constant    2.3829   0.02584  92.223   0.000
pH          0.4725   0.03418  13.824   0.000
Temp        0.3475   0.03418  10.167   0.001


S = 0.0683609   PRESS = 0.0485621
R-Sq = 98.66%   R-Sq(pred) = 96.52%   R-Sq(adj) = 97.99%


Analysis of Variance for Rate

Source           DF   Seq SS    Adj SS    Adj MS       F       P
Regression        2  1.37605   1.37605  0.688025  147.23   0.000
  Linear          2  1.37605   1.37605  0.688025  147.23   0.000
    pH            1  0.89302   0.89302  0.893025  191.09   0.000
    Temp          1  0.48302   0.48302  0.483025  103.36   0.001
Residual Error    4  0.01869   0.01869  0.004673
  Lack-of-Fit     2  0.00563   0.00563  0.002813    0.43   0.699
  Pure Error      2  0.01307   0.01307  0.006533
Total             6  1.39474
```

The fit of the linear model is good. We are most probably far from the maximum.

The model that locally best fits the surface is:

$$Y = 2.38 + 0.47\, X_1 + 0.35\, X_2 \qquad \text{(in coded units)}$$

X_1 is the *pH* value and X_2 is the temperature, in coded units.
The formula to go from uncoded units to coded units is:

$$X_1 = \frac{pH - \dfrac{3+4}{2}}{\dfrac{4-3}{2}} = \frac{pH - 3.5}{0.5} \qquad X_2 = \frac{Temp - \dfrac{20+25}{2}}{\dfrac{25-20}{2}} = \frac{Temp - 22.5}{2.5}$$

Hence, we go from coded to uncoded units by:

$$pH = 0.5\, X_1 + 3.5 \qquad Temp = 2.5\, X_2 + 22.5$$

This is the contour plot that corresponds to the model we have found:

Stat > DOE > Response Surface > Contour/Surface Plots

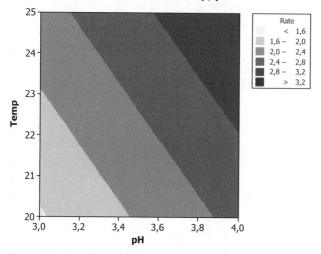

The direction of the maximum growth (steepest ascent) is determined by the vector

$$\left(\frac{\partial Y}{\partial X_1}, \frac{\partial Y}{\partial X_2} \right).$$

Remember that our model is $Y = 2.38 + 0.47\,X_1 + 0.35\,X_2$. Hence, in our case the direction of the steepest ascent is given by the vector (0.47; 0.35). The normalized vector is $u = (0.8; 0.6)$.

→ Step 2

In the second phase, we will make experiments in the direction of the steepest ascent.

| | Coded units | Uncoded units | | |
	(X1; X2)	pH	Temp	Rate
3u	(2.4; 1.8)	4.7	27.0	$4.26 \cdot 10^{-2}$
5u	(4.0; 3.0)	5.5	30.0	$5.59 \cdot 10^{-2}$
7u	(5.6; 4.2)	6.3	33.0	$6.32 \cdot 10^{-2}$
9u	(7.2; 5.4)	7.1	36.0	$5.21 \cdot 10^{-2}$

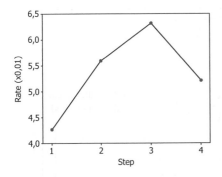

It seems that the maximum is around a pH value of 6.3 and a temperature of 33°C. In the third phase, we will perform a new 2^2 design with centre points to check whether we need quadratic terms to model this area of the surface.

→ Step 3

We choose the low and high levels according to the following table. Basically, the center point for pH and temperature are those conditions from the previous phase that provided the maximum growth rate (just the temperature changes slightly).

	Level −	Level +
pH	5,8	6,8
Temperature	30	35

The runs and the results obtained are:

pH	Temp [°C]	Tasa [$\cdot 10^{-2}$ min^{-1}]
5.8	30.0	5.02
6.8	30.0	5.51
5.8	35.0	4.54
6.8	35.0	4.89
6.3	32.5	6.16
6.3	32.5	6.23
6.3	32.5	6.10

Just as we have done in step 1, we fit a linear model to the data to see if it fits well. The result is the following:

```
Response Surface Regression: Rate versus pH; Temp

The analysis was done using coded units.

Estimated Regression Coefficients for Rate

Term          Coef   SE Coef       T       P
Constant    5.4169    2.6665   2.031   0.112          The fit of the linear
pH          0.2100    0.3851   0.545   0.615          model is not good.
Temp       -0.2750    0.3851  -0.714   0.515


S = 0.770299    PRESS = 9.81541
R-Sq = 16.79%   R-Sq(pred) = 0.00%   R-Sq(adj) = 0.00%


Analysis of Variance for Rate

Source           DF   Seq SS    Adj SS    Adj MS      F       P
Regression        2  0.47890   0.47890   0.23945   0.40   0.692
  Linear          2  0.47890   0.47890   0.23945   0.40   0.692
    pH            1  0.17640   0.17640   0.17640   0.30   0.615
    Temp          1  0.30250   0.30250   0.30250   0.51   0.515
Residual Error    4  2.37344   2.37344   0.59336
  Lack-of-Fit     2  2.36498   2.36498   1.18249  279.33  (0.004)
  Pure Error      2  0.00847   0.00847   0.00423
Total             6  2.85234
```

The fit of the linear model is not good: we probably need to include quadratic terms in the model. To be able to estimate these quadratic terms, we need to carry out more experiments.

→ Step 4

We add the points of the star to the experiments of step 3 (which corresponded to the points of the cube, including the center points). This is the complete matrix of the central composite design:

Block	pH	Temp [°C]	Rate [$\cdot 10^{-2}$ min^{-1}]
1	5.8	30.0	5.02
1	6.8	30.0	5.51
1	5.8	35.0	4.54
1	6.8	35.0	4.89
1	6.3	32.5	6.16
1	6.3	32.5	6.23
1	6.3	32.5	6.10
2	5.6	32.5	4.82
2	7.0	32.5	5.67
2	6.3	29.0	5.44
2	6.3	36.0	4.54
2	6.3	32.5	6.28
2	6.3	32.5	6.30
2	6.3	32.5	6.01

Block 1 corresponds to the experiments of step 3 (cube), block 2 to the experiments of step 4 (star). Since time has passed between the experiments of step 3 and those of step 4, it is a good idea to incorporate the block in the model to be able to estimate the effects free from the possible block effect.

Stat > DOE > Response Surface > Define Custom Response Surface Design

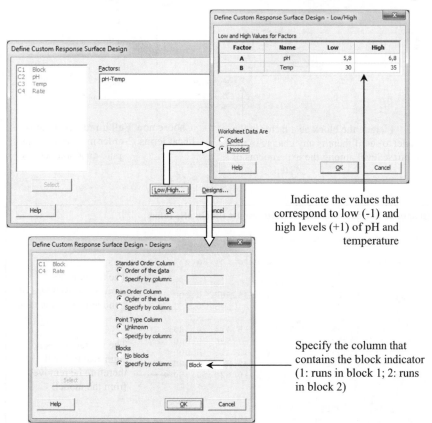

Indicate the values that correspond to low (-1) and high levels (+1) of pH and temperature

Specify the column that contains the block indicator (1: runs in block 1; 2: runs in block 2)

Stat > DOE > Response Surface > Analyze Response Surface Design

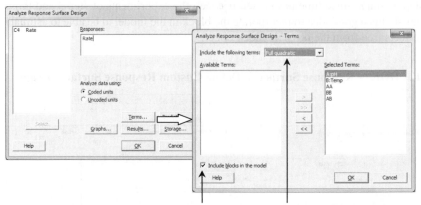

Include the block as a factor in the model to see if there is any change in the response level among the experiments of steps 3 and 4

Choose now **Full quadratic** (that is, linear terms, 2-order interactions and pure quadratic terms)

Response Surface Regression: Rate versus Block; pH; Temp

The analysis was done using coded units.

Estimated Regression Coefficients for Rate

Term	Coef	SE Coef	T	P
Constant	6.18021	0.04549	135.867	0.000
Block	-0.03707	0.02978	-1.245	0.253
pH	0.25631	0.03959	6.474	0.000
Temp	-0.29798	0.03959	-7.526	0.000
pH*pH	-0.50394	0.04161	-12.111	0.000
Temp*Temp	-0.63405	0.04161	-15.237	0.000
pH*Temp	-0.03500	0.05571	-0.628	0.550

The block is not significant; we therefore remove it from the model

The interaction is not significant; it will therefore be removed from the model.

S = 0.111423   PRESS = 0.363666
R-Sq = 98.49%   R-Sq(pred) = 93.69%   R-Sq(adj) = 97.20%

Analysis of Variance for Rate

Source	DF	Seq SS	Adj SS	Adj MS	F	P
Blocks	1	0.02658	0.01923	0.01923	1.55	0.253
Regression	5	5.64564	5.64564	1.12913	90.95	0.000
Linear	2	1.22355	1.22355	0.61177	49.28	0.000
pH	1	0.52032	0.52032	0.52032	41.91	0.000
Temp	1	0.70323	0.70323	0.70323	56.64	0.000
Square	2	4.41719	4.41719	2.20859	177.89	0.000
pH*pH	1	1.53467	1.82094	1.82094	146.67	0.000
Temp*Temp	1	2.88252	2.88252	2.88252	232.18	0.000

```
   Interaction     1   0.00490   0.00490   0.00490    0.39   0.550
      pH*Temp      1   0.00490   0.00490   0.00490    0.39   0.550
Residual Error     7   0.08691   0.08691   0.01242
   Lack-of-Fit     3   0.02597   0.02597   0.00866    0.57   0.665  ←┐
   Pure Error      4   0.06093   0.06093   0.01523
Total             13   5.75912
```

The model fit
is satisfactory

```
Unusual Observations for Rate

Obs  StdOrder  Rate    Fit   SE Fit  Residual  St Resid
 14        14  6.010  6.217   0.054   -0.207     -2.13 R

R denotes an observation with a large standardized residual.

...
```

We remove the block factor and the interaction to obtain the final model:

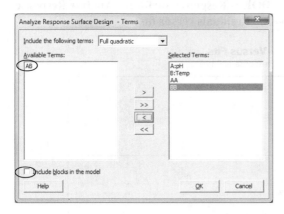

Response Surface Regression: Rate versus pH; Temp

```
The analysis was done using coded units.
Estimated Regression Coefficients for Rate

Term           Coef   SE Coef        T       P
Constant     6.1807   0.04534  136.309   0.000
pH           0.2563   0.03947    6.494   0.000
Temp        -0.2980   0.03947   -7.550   0.000
pH*pH       -0.5044   0.04148  -12.160   0.000
Temp*Temp   -0.6345   0.04148  -15.296   0.000


S = 0.111076    PRESS = 0.311694
R-Sq = 98.07%   R-Sq(pred) = 94.59%   R-Sq(adj) = 97.21%
```

```
Analysis of Variance for Rate

Source           DF    Seq SS    Adj SS    Adj MS       F      P
Regression        4   5.64808   5.64808   1.41202  114.45  0.000
  Linear          2   1.22355   1.22355   0.61177   49.59  0.000
    pH            1   0.52032   0.52032   0.52032   42.17  0.000
    Temp          1   0.70323   0.70323   0.70323   57.00  0.000
  Square          2   4.42453   4.42453   2.21227  179.31  0.000
    pH*pH         1   1.53780   1.82426   1.82426  147.86  0.000
    Temp*Temp     1   2.88674   2.88674   2.88674  233.97  0.000
Residual Error    9   0.11104   0.11104   0.01234
  Lack-of-Fit     4   0.04844   0.04844   0.01211    0.97  0.499
  Pure Error      5   0.06260   0.06260   0.01252
Total            13   5.75912
```

We can have a look at the graph of the residuals versus fitted values, which does not show anything special, by: **Stat > DOE > Response Surface > Analyze Response Surface Design > Graphs** (choosing **Residuals versus fits**)

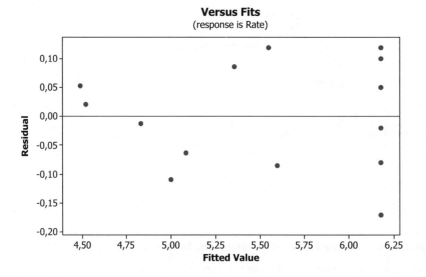

Versus Fits
(response is Rate)

The final model in the area of the maximum is:

$$Y = 6.18 + 0.26\, X_1 - 0.30\, X_2 - 0.50\, X_1^2 - 0.63\, X_2^2 \qquad \text{(in coded units)}$$

X_1 is the pH value and X_2 is the temperature, in codified units.

The contour plot and the response surface can be drawn by **Stat > DOE > Response Surface > Contour/Surface Plots**

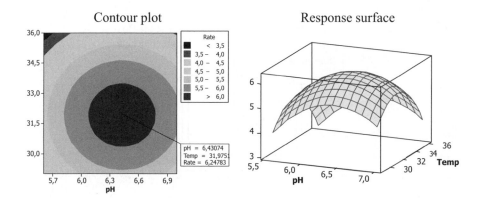

Contour plot   Response surface

It can be seen in the contour plot that the maximum is obtained at a pH value of 6.4 and a temperature of 32 °C.

30.6   Jam

A food company decides to improve the taste of one of the jams it produces. To do so, different prototypes of the jam are prepared by varying the content in grams of three ingredients of the formula. Each prototype is then given to different consumers who grade the taste using a questionnaire. The overall score of each prototype is an average of the answers to the questionnaire's questions.

The 3 factors involved in the experiment, together with their low and high levels, are shown in the following table.

	Level −	Level +
Sugar (S)	2.0	4.0
Ginger (G)	0.5	1.0
Mango (M)	4.0	8.0

The objective is to decide which prototypes will be used and analyze the results once the experiment has been done. There is a restriction that must be taken into account in the design of the experiment: it is not possible to experiment with all factors at the high level or all factors at the low level.

At this stage of the study, we want to perform an experiment that makes it possible to obtain a model with purely quadratic terms. One can use the Box-Behnken design which has the advantage of never experimenting with all factors at either high or low levels.

Stat > DOE > Response Surface > Create Response Surface Design

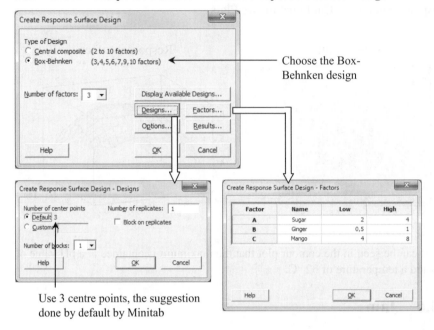

Choose the Box-Behnken design

Use 3 centre points, the suggestion done by default by Minitab

We carry out the experiments in random order as proposed by Minitab. The following table shows the answers for each of the experimental conditions:

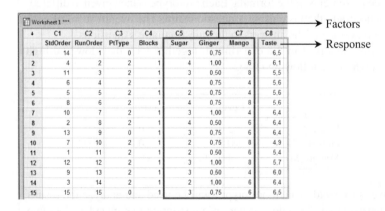

Factors

Response

If you want to reproduce this example, have in mind that the runs are not in standard order.

Analyze the experiment: **Stat > DOE > Response Surface > Analyze Response Surface Design**

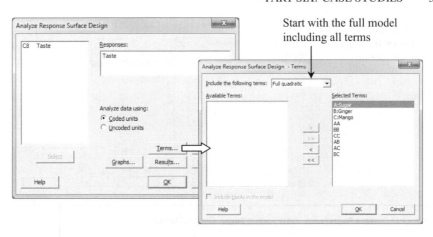

Start with the full model
including all terms

Response Surface Regression: Taste versus Sugar; Ginger; Mango

The analysis was done using coded units.

Estimated Regression Coefficients for Taste

Term	Coef	SE Coef	T	P
Constant	6.46667	0.05110	126.552	0.000
Sugar	0.17500	0.03129	5.593	0.003
Ginger	0.16250	0.03129	5.193	0.003
Mango	-0.23750	0.03129	-7.590	0.001
Sugar*Sugar	-0.43333	0.04606	-9.408	0.000
Ginger*Ginger	0.04167	0.04606	0.905	0.407 ←
Mango*Mango	-0.60833	0.04606	-13.207	0.000
Sugar*Ginger	-0.32500	0.04425	-7.344	0.001
Sugar*Mango	0.17500	0.04425	3.955	0.011
Ginger*Mango	-0.05000	0.04425	-1.130	0.310 ←

S = 0.0885061 PRESS = 0.535
R-Sq = 98.87% R-Sq(pred) = 84.60% R-Sq(adj) = 96.84%

Analysis of Variance for Taste

Source	DF	Seq SS	Adj SS	Adj MS	F	P
Regression	9	3.43417	3.43417	0.38157	48.71	0.000
Linear	3	0.90750	0.90750	0.30250	38.62	0.001
Sugar	1	0.24500	0.24500	0.24500	31.28	0.003
Ginger	1	0.21125	0.21125	0.21125	26.97	0.003
Mango	1	0.45125	0.45125	0.45125	57.61	0.001
Square	3	1.97167	1.97167	0.65722	83.90	0.000
Sugar*Sugar	1	0.57619	0.69333	0.69333	88.51	0.000
Ginger*Ginger	1	0.02907	0.00641	0.00641	0.82	0.407
Mango*Mango	1	1.36641	1.36641	1.36641	174.44	0.000
Interaction	3	0.55500	0.55500	0.18500	23.62	0.002
Sugar*Ginger	1	0.42250	0.42250	0.42250	53.94	0.001
Sugar*Mango	1	0.12250	0.12250	0.12250	15.64	0.011
Ginger*Mango	1	0.01000	0.01000	0.01000	1.28	0.310
Residual Error	5	0.03917	0.03917	0.00783		
Lack-of-Fit	3	0.03250	0.03250	0.01083	3.25	0.244
Pure Error	2	0.00667	0.00667	0.00333		
Total	14	3.47333				

Remove the nonsignificant quadratic term *Ginger²* and the interaction).

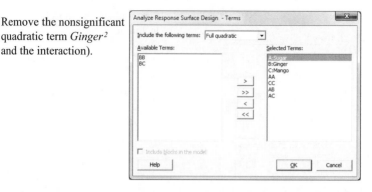

Response Surface Regression: Taste versus Sugar; Ginger; Mango

```
The analysis was done using coded units.
Estimated Regression Coefficients for Taste

Term            Coef   SE Coef        T      P
Constant      6.4923   0.04280  151.674  0.000
Sugar         0.1750   0.03150    5.555  0.001
Ginger        0.1625   0.03150    5.158  0.001
Mango        -0.2375   0.03150   -7.539  0.000
Sugar*Sugar  -0.4365   0.04623   -9.442  0.000
Mango*Mango  -0.6115   0.04623  -13.227  0.000
Sugar*Ginger -0.3250   0.04455   -7.295  0.000
Sugar*Mango   0.1750   0.04455    3.928  0.006


S = 0.0891042  PRESS = 0.306646
R-Sq = 98.40%  R-Sq(pred) = 91.17%  R-Sq(adj) = 96.80%


Analysis of Variance for Taste

Source           DF   Seq SS   Adj SS   Adj MS       F      P
Regression        7  3.41776  3.41776  0.48825   61.50  0.000
  Linear          3  0.90750  0.90750  0.30250   38.10  0.000
    Sugar         1  0.24500  0.24500  0.24500   30.86  0.001
    Ginger        1  0.21125  0.21125  0.21125   26.61  0.001
    Mango         1  0.45125  0.45125  0.45125   56.84  0.000
  Square          2  1.96526  1.96526  0.98263  123.76  0.000
    Sugar*Sugar   1  0.57619  0.70782  0.70782   89.15  0.000
    Mango*Mango   1  1.38907  1.38907  1.38907  174.96  0.000
  Interaction     2  0.54500  0.54500  0.27250   34.32  0.000
    Sugar*Ginger  1  0.42250  0.42250  0.42250   53.21  0.000
    Sugar*Mango   1  0.12250  0.12250  0.12250   15.43  0.006
Residual Error    7  0.05558  0.05558  0.00794
  Lack-of-Fit     5  0.04891  0.04891  0.00978    2.93  0.273
  Pure Error      2  0.00667  0.00667  0.00333
Total            14  3.47333
```
The model fits correctly ↑

Therefore, the final model, using coded units, is:

$$Y = 6.49 + 0.18X_1 + 0.16X_2 - 0.24X_3 - 0.44X_1^2 - 0.61X_2^2 - 0.33X_1X_2 + 0.18X_1X_2$$

Where X_1 is sugar, X_2 is ginger, and X_3 is mango.

To interpret the model, draw the contour plot setting X_2 (ginger) at its low, high and medium level (we fix X_2 because it is the only factor that does not appear in the model with its pure quadratic term. The quadratic terms X_1^2 and X_3^2 do appear in the model and we want to have X_1 and X_3 in the axes to see their curvature).

Stat > DOE > Response Surface > Contour/Surface Plots and choose Contour plot

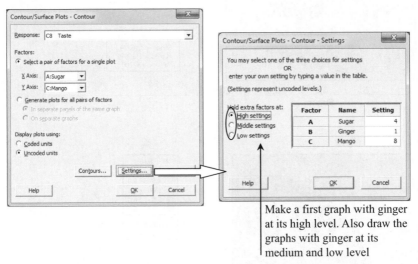

Make a first graph with ginger at its high level. Also draw the graphs with ginger at its medium and low level

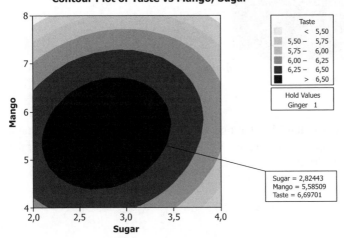

Contour Plot of Taste vs Mango; Sugar

Sugar = 2,82443
Mango = 5,58509
Taste = 6,69701

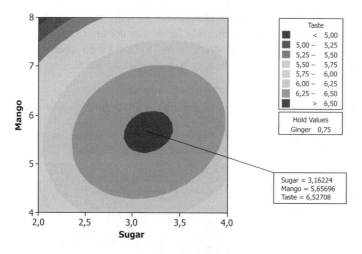

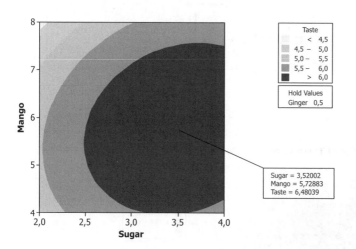

The maximum value is obtained in the first graph (Ginger: 1.0). It is in this graph where the 'mountain' reaches its highest value. As the ginger value decreases, the maximum moves to the right.

30.7   Photocopies

The following 35 data of file PHOTOCOPIES.MTW correspond to the duration (measured in number of copies) of toner cartridges of 35 photocopiers of the same type used in 35 offices:

13471	19053	17626	10748	7849	7271	648
22566	4211	63001	4078	8336	5199	1965
20812	2177	141	1527	65529	18247	50060
21935	27118	7889	12462	21554	4815	4944
9364	2169	47364	4580	794	6275	23710

The objective is to find out the probability that a toner cartridge continues operating after 17500 copies.

That is, we want to know its reliability after 17500 copies.

As a first analysis, we can try to answer the question by means of a nonparametric data analysis, without assuming any theoretical model.

Stat > Reliability/Survival > Distribution Analysis (Right Censoring) > Nonparametric Distribution Analysis

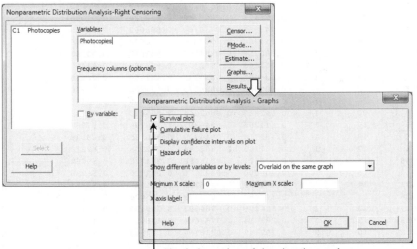

Check the option of showing the graph of the empirical reliability function

Distribution Analysis: Photocopies

Variable: Photocopies

Censoring Information   Count
Uncensored value   35

Nonparametric Estimates

Characteristics of Variable

	Standard	95,0%	Normal CI
Mean(MTTF)	Error	Lower	Upper
15413.9	2878.70	9771.80	21056.1

Median = 8336
IQR = 17343   Q1 = 4211   Q3 = 21554

Kaplan-Meier Estimates

Time	Number at Risk	Number Failed	Survival Probability	Standard Error	95,0% Lower	Normal CI Upper
141	35	1	0.971429	0.0281603	0.916235	1.00000
648	34	1	0.942857	0.0392347	0.865959	1.00000
794	33	1	0.914286	0.0473188	0.821543	1.00000
1527	32	1	0.885714	0.0537785	0.780310	0.99112
1965	31	1	0.857143	0.0591485	0.741214	0.97307
2169	30	1	0.828571	0.0637049	0.703712	0.95343
2177	29	1	0.800000	0.0676123	0.667482	0.93252
4078	28	1	0.771429	0.0709782	0.632314	0.91054
4211	27	1	0.742857	0.0738764	0.598062	0.88765
4580	26	1	0.714286	0.0763604	0.564622	0.86395
4815	25	1	0.685714	0.0784693	0.531917	0.83951
4944	24	1	0.657143	0.0802329	0.499889	0.81440
5199	23	1	0.628571	0.0816735	0.468494	0.78865
6275	22	1	0.600000	0.0828079	0.437700	0.76230
7271	21	1	0.571429	0.0836486	0.407480	0.73538
7849	20	1	0.542857	0.0842044	0.377820	0.70789
7889	19	1	0.514286	0.0844809	0.348706	0.67987
8336	18	1	0.485714	0.0844809	0.320135	0.65129
9364	17	1	0.457143	0.0842044	0.292105	0.62218
10748	16	1	0.428571	0.0836486	0.264623	0.59252
12462	15	1	0.400000	0.0828079	0.237700	0.56230
13471	**14**	**1**	**0.371429**	**0.0816735**	**0.211352**	**0.53151**
17626	13	1	0.342857	0.0802329	0.185604	0.50011
18247	12	1	0.314286	0.0784693	0.160489	0.46808
19053	11	1	0.285714	0.0763604	0.136051	0.43538
20812	10	1	0.257143	0.0738764	0.112348	0.40194
21554	9	1	0.228571	0.0709782	0.089457	0.36769
21935	8	1	0.200000	0.0676123	0.067482	0.33252
22566	7	1	0.171429	0.0637049	0.046569	0.29629
23710	6	1	0.142857	0.0591485	0.026928	0.25879
27118	5	1	0.114286	0.0537785	0.008882	0.21969
47364	4	1	0.085714	0.0473188	0.000000	0.17846
50060	3	1	0.057143	0.0392347	0.000000	0.13404
63001	2	1	0.028571	0.0281603	0.000000	0.08376
65529	1	1	0.000000	0.0000000	0.000000	0.00000

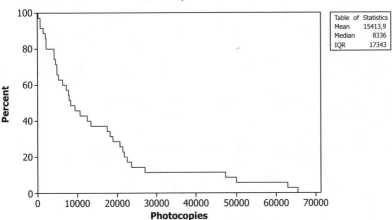

Survival Plot for Photocopies
Kaplan-Meier Method
Complete Data

We can determine the reliability after 17 500 copies looking at the row in bold face in the previous table. At 13 471 copies the reliability is 0.3714. No other toner fails until 17 500 copies and therefore, the empirical reliability at 17 500 copies is 0.3714.

We will now check which theoretical models best fit our data:

Stat > Reliability/Survival > Distribution Analysis (Right Censoring) > Distribution ID Plot

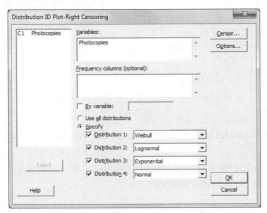

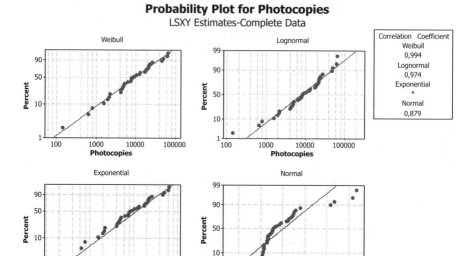

In this case, our data approximately form a straight line on the Weibull probability plot. In addition, the output in the session window shows us the Anderson-Darling statistic and the correlation coefficient for each model.

Distribution ID Plot: Photocopies

Goodness-of-Fit

Distribution	Anderson-Darling (adj)	Correlation Coefficient
Weibull	0.596	0.994
Lognormal	0.764	0.974
Exponential	0.811	*
Normal	3.153	0.879

This output confirms that the Weibull model is the parametric model that best fits the data.

Using the Weibull model, we can now estimate the probability that a cartridge exceeds 17 500 copies by means of a parametric analysis.

Stat > Reliability/Survival > Distribution Analysis (Right Censoring) > Parametric Distribution Analysis

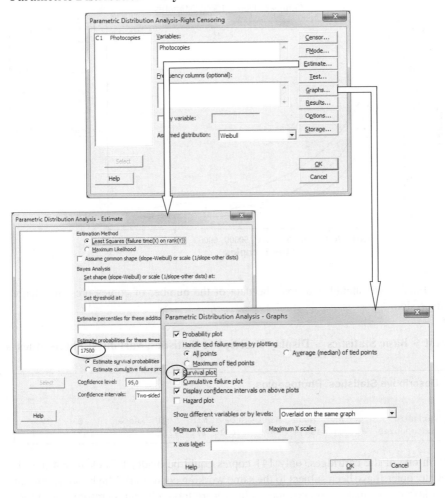

Table of Survival Probabilities

		95,0%	Normal CI
Time	Probability	Lower	Upper
17500	0.313499	0.196119	0.437818

The reliability at 17 500 copies (that corresponds to the probability that the cartridge exceeds 17 500 copies) is 0.3135. Remember that the nonparametric estimation of the reliability was 0.3714.

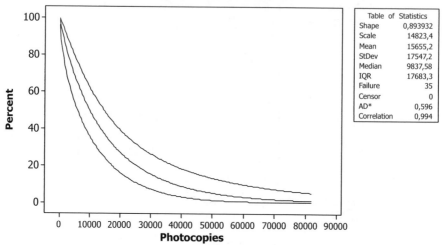

Survival Plot for Photocopies
Weibull - 95% CI
Complete Data - LSXY Estimates

Table of Statistics	
Shape	0,893932
Scale	14823,4
Mean	15655,2
StDev	17547,2
Median	9837,58
IQR	17683,3
Failure	35
Censor	0
AD*	0,596
Correlation	0,994

Finally, we should note that the data of the number of copies present a large dispersion.

Stat > Basic Statistics > Display Descriptive Statistics, Variables: Photocopies

Descriptive Statistics: Photocopies

Variable	N	N*	SE Mean	Mean	StDev	Min	Q1	Median	Q3	Max
Photocopies	35	0	15414	2879	17031	141	4211	8336	21554	65529

With one of the cartridges, only 141 copies could be made! It is clear that not all photocopiers have been subject to the same working conditions. Maybe some did not work too well, making photocopies which were too dark and therefore used more ink. Or it is simply possible that in some offices, copies were made that needed much more ink per sheet?

The test is not badly designed if what we want is to capture the whole variability caused by the use of the photocopiers in different offices. However, if the aim was to study the duration of the toner in 'laboratory conditions', a test would have had to be designed that specified, among other aspects, the amount of 'black' on the sheets to be photocopied.

APPENDICES

1. **Answers to Questions that Arise at the Beginning**
2. **Managing Data**
3. **Customization of Minitab**

APPENDICES

A1

Appendix 1: Answers to Questions that Arise at the Beginning

1. Why do some columns that should contain data appear to be empty?

It is possible that you do not see the data because they are located on the first rows, which are not visible.

↓	C1	C2	C3	C4	C5	C6	C7
13	7	15	11				
14	19	14	12				
15	5	11	12				
16	19	9	6				
17	9	7	17				
18	9	14	15				

Worksheet 1 ***

Each of the columns C4, C5 and C6 contain 10 values. These first 10 values are not seen because the worksheet only shows data values from row 13.

2. I cannot make any operation with the data contained in a column. What is happening?

Most probably the column data is in text format. A column is automatically assigned a text format when any cell contains text or a symbol different from * (missing value symbol).

When a column is in text format, it does not automatically change to a numeric format even though the text or symbols contained in the column are eliminated.

Industrial Statistics with Minitab, First Edition. Pere Grima Cintas, Lluís Marco-Almagro and Xavier Tort-Martorell Llabrés.
© 2012 John Wiley & Sons, Ltd. Published 2012 by John Wiley & Sons, Ltd.

3. I saw some instructions in the menus and now there is no way to find them. What is going on?

The content of the menus depends on which is the active window (session, worksheet, graphical window...). Click on another Minitab window if the command that you are looking for does not appear in a menu.

Menu **Editor** with session window active   Menu **Editor** with worksheet active

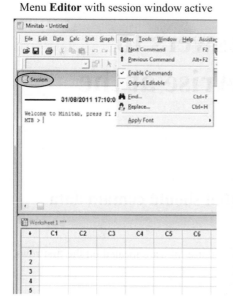

4. All menu options are inactive. How can I return to a normal situation?

Probably, you are in the middle of writing a line in the session window. To finish the command line press [Enter]. To end a line that asks for a subcommand, type a period and press [Enter].

Some actions, such as the stepwise regression, finish with a question. That question has to be answered to make the menus active and to continue.

5. I have a column containing numbers, but it appears with a text format. What should I do?

Probably, there is a cell in the column containing a letter or a symbol (or there has been one at some point in the past). You can change the data type to a numeric format doing:

Data > Change Data Type > Text to Numeric

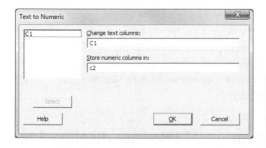

The cells that contain no numeric values are transformed to missing values

Observe that in text columns a T appears next to the column and their values are aligned to the left. Numeric columns, however, are aligned to the right.

6. I am creating a graph and it is displayed with options that I have not chosen. Why?

A Minitab dialog box keeps the options and values specified in the last access to that dialog box in the same session. These options and values are remembered until they are changed or the program is exited. Additionally, if you have stored your work as a Minitab project (*.MPJ), the dialog boxes will show the options they had when the project was saved.

If a graph does not have the expected appearance, enter again in the dialog box (remember that you can use [Ctrl] + [E]) and investigate dialog boxes with options that can be accessed from the main dialog box. Some option has most likely been modified in a previous analysis, and now you can set it back to its default value.

7. I have stored a worksheet but I do not know where it is located. What can I do?

Files are stored in the directory that has been defined using: **Tools > Options**

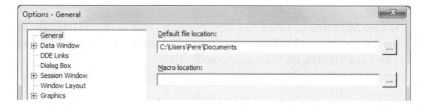

However, in case the default directory is changed at the moment of saving a file, all files used in the actual working session (while you do not exit Minitab) will be stored in the new directory. To find out where files are being stored you can proceed to save the actual worksheet and observe what directory is being used.

Another possible situation is that perhaps you saved the file as a Project (MPJ extension) – which happens when clicking on the Save icon – and you are trying to open it as a worksheet (MTW extension) or vice versa. Make a search with *.* so all files contained in the directory are displayed.

8.   Why is the Minitab prompt MTB> not displayed in the session window?

Make the session window active and then do:

Editor > Enable Commands

To make the session window active, click over any point on this window (the color of the frame will become darker).

Note that this action must be repeated every time you enter Minitab. However, you can configure it to always display the prompt **MTB >**. To achieve this, choose: **Tools > Options: Session Window** (click on the cross on the left) > **Submitting Commands:** choose the **Enable** option under **Command Language**.

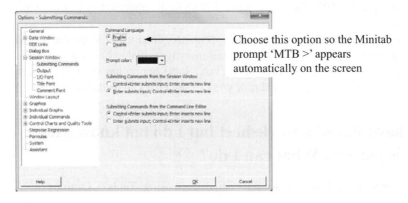

Choose this option so the Minitab prompt 'MTB >' appears automatically on the screen

In case Minitab is installed on a server, you can only configure it to always display the prompt **MTB >** if you have write permissions in the directory where the Minitab application is stored.

A2

Appendix 2: Managing Data

In many instances, managing the data to have it in an appropriate manner to carry out statistical analysis is a laborious task; sometimes more so than the analysis itself. Thus, it is important to develop skills in managing data and this is mainly achieved with practice, facing situations that once solved contribute to expand our baggage of resources – and even tricks – to use.

This appendix provides a general idea, using examples, of the three most common activities for managing data:

– Copying columns with restrictions

– Stacking and unstacking columns

– Codifying and sorting data

A2.1 Copy Columns with Restrictions (File: 'PULSE')

 Consider again the file PULSE.MTW, used in Chapter 2. Remember that its content is collected in a class with 92 students, where initially each student registers its own height, weight, gender, smoking preference, usual physical activity level and resting pulse rate. Then, each student flips a coin and those who get a face coming up must run in place during one minute. Finally, after that time, all students measure and register their new pulse rates. The following columns of the worksheet will be used:

Industrial Statistics with Minitab, First Edition. Pere Grima Cintas, Lluís Marco-Almagro and Xavier Tort-Martorell Llabrés.
© 2012 John Wiley & Sons, Ltd. Published 2012 by John Wiley & Sons, Ltd.

Column	Name	Contents
C1	Pulse1	Initial pulse of the 92 students
C2	Pulse2	Final pulse
C3	Ran	1=ran in place; 2=did not run in place
C5	Sex	1=male; 2=female

The aim is to compare the pulse increment between males and females who have run. This can be achieved using some useful data management options.

Calculate first the difference between Pulse1 and Pulse2 using **Calculator**.

Calc > Calculator

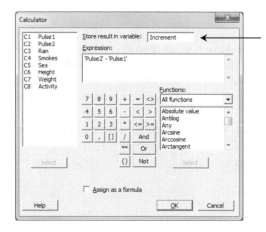

Assign a name to the new column. Minitab places it in the first empty column

One might think that the comparison of the increments stratified by gender could be done now, but this is not so, because the increment variable includes people who have run and people who have not. To place in a column only the increments of those who have run, a copy of selected values from the column containing the increments is made.

Data > Copy > Columns to Columns

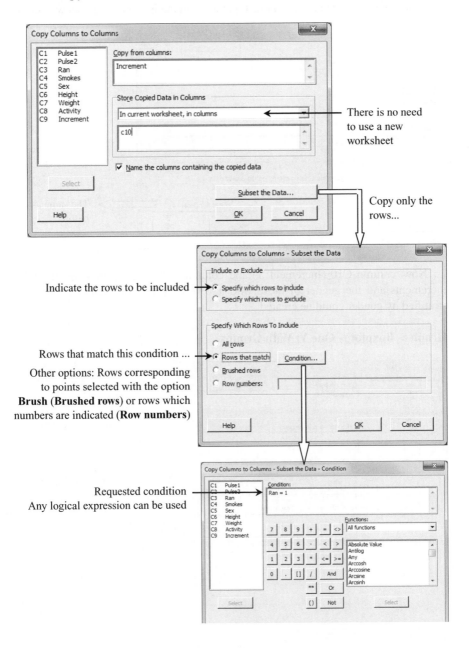

Now, column C10 contains the pulse increments of the students who have run. But these cannot be stratified, because in column C5 we have the gender of all students, both those that have run and those that have not. We must copy in a new column the sex of those that ran. This can be done together with the increments, as shown below:

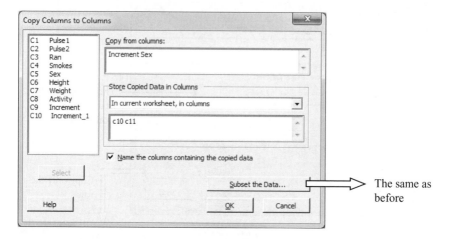

Now, column C10 (Increment_1) and column C11 (Sex_1) contains the pulse increments and the gender of those who have run. Then, for example, a boxplot stratified by gender could be created.

Graph > Boxplot > One Y: With Groups

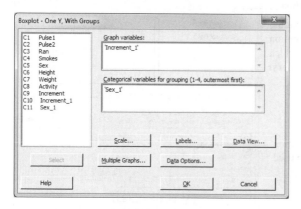

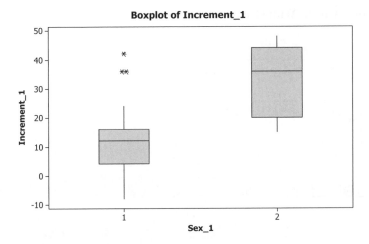

The increment is higher in women (code 2) than in men (code 1): the first quartile in women is above the third quartile of men. There are three men with anomalous increments, but if these values were attained by women they would be inside the interquartile range. It is surprising to find men with negative increment values; this casts some doubts on the rigor with which data were taken.

A2.2 Selection of Data when Plotting a Graph

The same result obtained above can be produced by selecting the data directly, in the same moment that the graph is being created, through the graphic button: **Data Options**.

Graph > Boxplot > One Y: With Groups

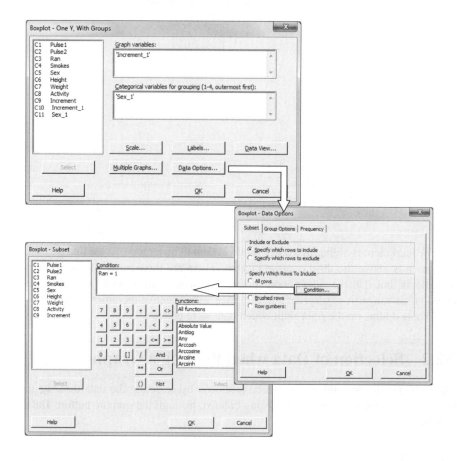

The obtained result is exactly the same as the one obtained before, although the used values do not remain stored separately in the worksheet.

A2.3   Stacking and Unstacking of Columns (File 'BREAD')

Consider again file BREAD.MTW already analyzed in chapter six. A detailed description of the steps needed to organize the data in the requested format is given.

The owner of a bakery has detected that some loafs of bread seem to have much less weight than expected. In the bakery, a loaf of bread is produced by two workmen (A and B) using two machines (M1 and M2). The operators do not work simultaneously, but in different days: some days the loafs of bread are produced by operator A using both machines, and some other days by operator B.

The owner of the bakery decides to collect data and carry out an analysis to find out what is causing the problems. For a period of 20 days (10 for operator A and 10 for operator B), four loafs of bread produced by machine one and four produced by machine two are taken at random. The obtained results, included in file BREAD.MTW, are organized as follows:

Column	Contents
C1	Day
C2	Operator
C3	Machine 1 part 1
C4	Machine 1 part 2
C5	Machine 1 part 3
C6	Machine 1 part 4
C7	Machine 2 part 1
C8	Machine 2 part 2
C9	Machine 2 part 3
C10	Machine 2 part 4

The nominal weight for the pieces of bread is 210 g, with a tolerable variation of $\pm$ 10 g. The aim is to analyze the collected data to identify what is causing problems.

First, create a column containing the weights of the loafs of bread produced by machine 1, and another column containing the weights produced by machine 2. This can be achieved using the option **Stack.**

Data > Stack > Columns

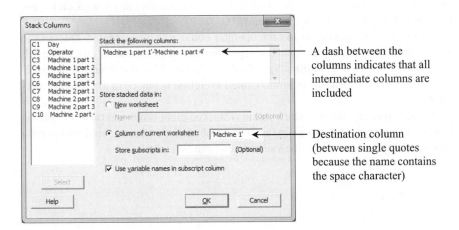

A dash between the columns indicates that all intermediate columns are included

Destination column (between single quotes because the name contains the space character)

Now, the weights produced by machine 1 are contained in a single column. Likewise, the weights produced by machine 2 can be placed in another column.

 If the name of the destination column (or any other) includes space characters, it should be written between single quotes; for example: 'Machine 1'.

After constructing a histogram for the weight data stratified by machines, it is clearly observed that the first machine is not centred.

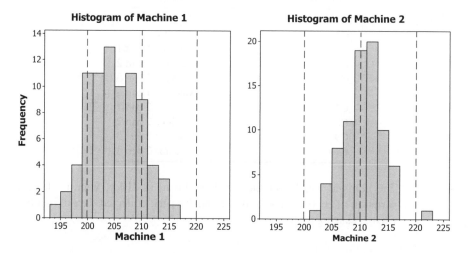

The default appearance of the histograms has been modified by changing the scales (forced to be equal to facilitate the comparison). Also, reference lines for the nominal values and the specification limits have been added (with option **Editor >Annotation > Graph Annotation Tools**).

Additionally, the default width and height values have been changed (right-click on the graphical window, but outside the frame containing the plot) using **Edit Figure Region > Graph Size**: Custom 100 and 100.

The values corresponding to each operator could also be compared. One way to achieve this is by stacking the data corresponding to machines 1 and 2 in a single column: **Data > Stack > Columns.** Specifically, columns 'Machine 1' and 'Machine 2' are stacked and the resulting column (named **Weights**) is placed in the current worksheet.

In the following, the operator column is stacked eight times and the result placed in a column named 'Operators'. In this way we obtain information about the operator corresponding to each one of the values in column 'Weights'.

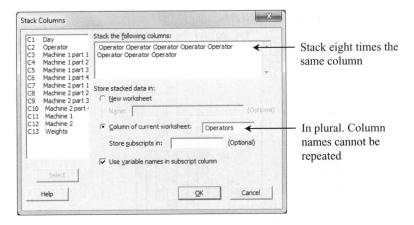

Stack eight times the same column

In plural. Column names cannot be repeated

Then, histograms for the weights stratified by operator are created and displayed in a single panel: **Graphs > Histogram > Simple**

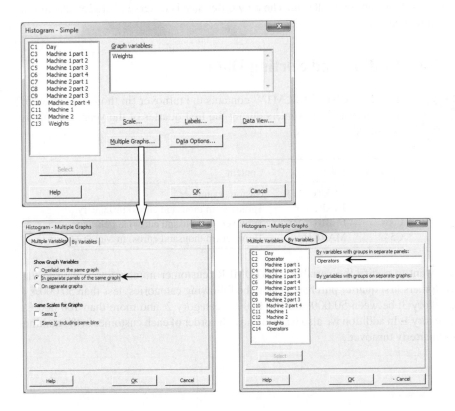

Histogram of Weights

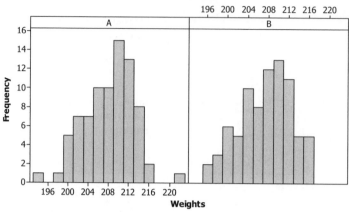

Panel variable: Operators

Stratification could also be performed by operator and machine, but the resulting analysis shows nothing relevant. The only difference is observed between machines, as aforementioned.

A2.4   Coding and Sorting Data

The file CUSTOMER.MTW contains the turnover (in thousands of euro) of a company's customers during the first quarter of the year. This file is organized as follows:

Column	Label	Content
C1	CUSTOMER	Customer number
C2	JANUARY	Turnover (in thousand euros) in January
C3	FEBRUARY	Turnover (in thousand euros) in February
C5	MARCH	Turnover (in thousand euros) in March

We want to codify the variable CUSTOMER (customer number) according to the customers first quarter purchases using the following categories: less than 50 000 € : category 3; between 50 000 and 100 000 € : category 2, and more than 100 000 € category 1. In addition we also want to know the order of each customer in the ranking of quarterly turnover.

First, calculate the turnover during the first quarter of the year.
Calc > Row Statistics

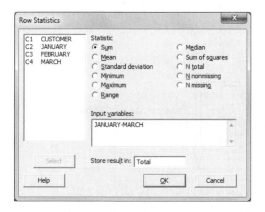

Then, the total turnover values are codified according to the previously specified categories: **Data > Code > Numeric to Numeric**

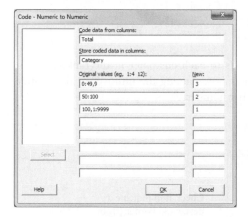

In this dialog box symbols like 'greater than (>)' or 'less than (<)' cannot be used. Intervals should be written in the indicated format

The customer numbers for each category can be placed in separate columns, doing: **Data > Unstack Columns**

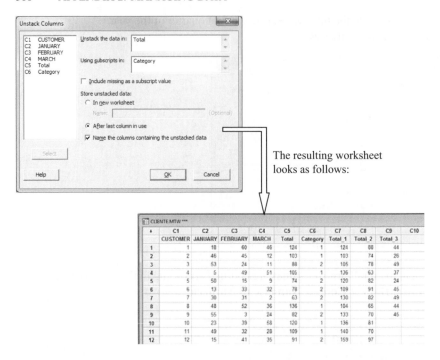

The resulting worksheet
looks as follows:

	C1	C2	C3	C4	C5	C6	C7	C8	C9	C10
	CUSTOMER	JANUARY	FEBRUARY	MARCH	Total	Category	Total_1	Total_2	Total_3	
1	1	18	60	46	124	1	124	88	44	
2	2	46	45	12	103	1	103	74	26	
3	3	63	24	11	88	2	105	78	49	
4	4	5	49	51	105	1	136	63	37	
5	5	50	15	9	74	2	120	82	24	
6	6	13	33	32	78	2	109	91	45	
7	7	30	31	2	63	2	130	82	49	
8	8	48	52	36	136	1	104	65	44	
9	9	55	3	24	82	2	133	70	45	
10	10	23	39	68	120	1	136	81		
11	11	49	32	28	109	1	140	70		
12	12	15	41	35	91	2	159	97		

To find out the order that corresponds to each customer in the ranking of the quarterly turnover, you can sort the column containing the turnover of the quarter, dragging in this arrangement the customer numbers.

Data > Sort

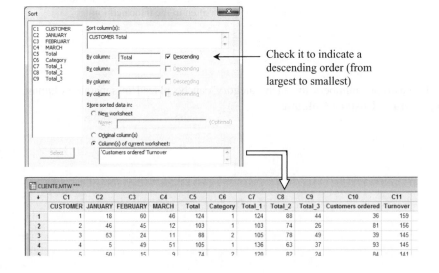

Check it to indicate a
descending order (from
largest to smallest)

	C1	C2	C3	C4	C5	C6	C7	C8	C9	C10	C11
	CUSTOMER	JANUARY	FEBRUARY	MARCH	Total	Category	Total_1	Total_2	Total_3	Customers ordered	Turnover
1	1	18	60	46	124	1	124	88	44	36	159
2	2	46	45	12	103	1	103	74	26	81	156
3	3	63	24	11	88	2	105	78	49	39	145
4	4	5	49	51	105	1	136	63	37	93	145
5	5	50	15	9	74	2	120	82	24	84	141

The customer number 36 is the first one in the ranking with a turnover value of 159 000 € . The customer number 81 is the second, etc. An alternative way to obtain the ranking is through the option:

Data > Rank

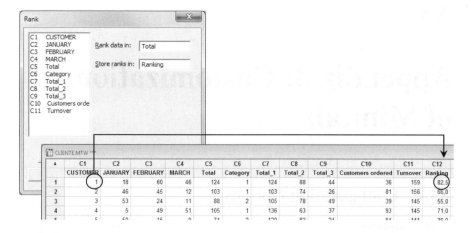

The ranking number is generated from the smallest to the largest turnover. Customer number 1 has rank 82.5. This is not an integer number because there are two customers with a turnover of 124 000 € . One would have rank 82 and the other rank 83, but Minitab assigns rank 82.5 to both.

A3

Appendix 3: Customization of Minitab

A3.1   Configuration Options

Tools > Options

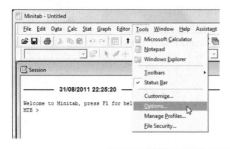

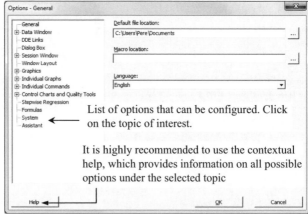

List of options that can be configured. Click on the topic of interest.

It is highly recommended to use the contextual help, which provides information on all possible options under the selected topic

Industrial Statistics with Minitab, First Edition. Pere Grima Cintas, Lluís Marco-Almagro and Xavier Tort-Martorell Llabrés.
© 2012 John Wiley & Sons, Ltd. Published 2012 by John Wiley & Sons, Ltd.

If you click on a topic with a + sign on the left, general information about this topic is displayed on the right side. If you click on the + sign, subtopics are shown.

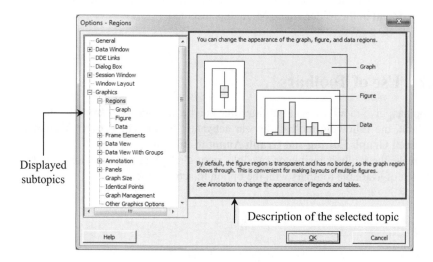

Some configuration options that could be relevant to change:

- **Session Window** > **Submitting Commands** and choose **Enable** for **Command Language**. Doing so, the Minitab prompt (symbol: MTB>) appears and instructions can be entered directly in the session window. Additionally, the instructions that are used through the menus are automatically displayed on the session window, which is very useful to write macros. More information on how to use this feature can be found at the book's website.

- The light brown background with a thin black border, appearing by default in Minitab graphs, may not be appropriate for plots in a document or in a report. Both the background and the border can be removed from **Graphics** > **Regions** > **Graph**. In **Fill Pattern Type**, choose N inside the box (**None**) and in **Borders and Fill Lines: Type**, choose **None**.

- It can be annoying that every time a graph is closed, Minitab asks if it should be saved. This can be disabled from **Graphs** > **Graph Management**. In **Prompt to Save a Graph Before Closing**, select **Never**.

There are many options for specific graphs. Most of them can be handled by double-clicking on the graph of interest. You can make that option the configuration by default for all graphs of this kind by changing it in **Tools** > **Options**.

The changes done with **Tools > Options** are stored in the Windows registry, in such a way that if you uninstall and reinstall Minitab, the options that you have modified will be preserved. If you want to reset Minitab to the default values, execute the file rmd.exe (Restore Minitab Defaults) that is found in the folder where Minitab is installed.

A3.2   Use of Toolbars

The toolbars give rapid access to utilities that can be also accessed from the menus. By default, the following toolbars appear activated: **Standard**, **Project Manager, Worksheet, Graph Editing** and **Graph Annotation Tools**.

To make the toolbars visible or invisible, right-click on any of them and you will get the list of available toolbars with the option to activate them.

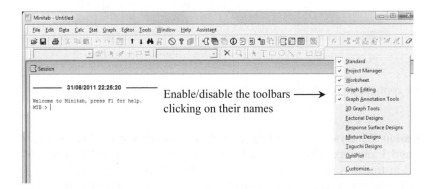

You can also display the list of available toolbars from **Tools > Toolbars**.

For instance, the location of the toolbars can be changed; they can be detached from the top menu and made into floating objects, or the toolbar size can be changed. Everything can be applied in ways typical to Windows user interfaces.

The toolbars commands are disabled when using them makes no sense. Some toolbars, such as the ones related to the graphs edition, may disappear when graphs are closed if they are not anchored to the top menu.

A3.3   Add Elements to an Existing Toolbar

In case some Minitab menu options are frequently used, you can add them to an existing toolbar.

Tools > **Customize**, or right-click on any toolbar, and choose **Customize** on the displayed menu.

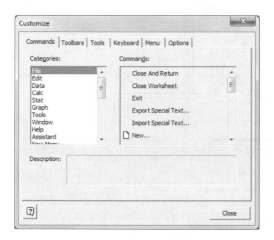

From the **Commands** tab, a list with all Minitab menus can be accessed. Any option from the right menu can be selected and then placed in an existing toolbar. The procedure to follow is commonly used in many Windows programs that allow you to customize toolbars.

To remove an option from a toolbar, simply pull it out by dragging it while the **Customize** window remains open.

 In the same way that options are added to a toolbar, they can also be added to a menu. With the **Customize** window open, drag the option of interest to the menu, and release it wherever you want it to stay. You can even drag commands inside the menus to change their location.

A3.4   Create Custom Toolbars

Imagine that the tools for comparing two treatments are frequently used. In the following, a toolbar is created with these options.

Tools > Customize, and click on the second tab, **Toolbars**, and then in **New**...

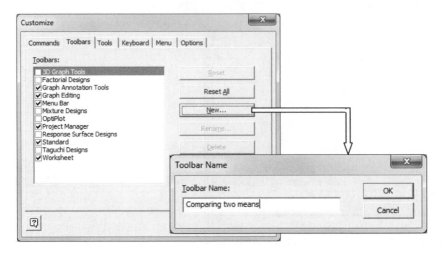

An empty toolbar appears that can be filled up with menu options as seen above.

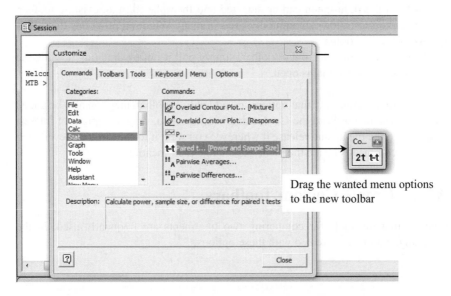

The options can be directly selected from the Minitab menus and then dragged to the toolbars. In that case, you should press the [Ctrl] key before releasing it, because on the contrary the command would be moved instead of being copied.

Before releasing the command over a toolbar, press the [Ctrl] key. The + sign appears next to the cursor, indicating that the created button is a copy of the menu option. If the [Ctrl] key is not pressed, the option will disappear from the menu (though if that occurs, it is not so serious, since from **Customize > Commands** you can drag the instructions not only to toolbars, but also to menus).

Index

Industrial Statistics with Minitab, First Edition. Pere Grima Cintas, Lluís Marco-Almagro and Xavier Tort-Martorell Llabrés.
© 2012 John Wiley & Sons, Ltd. Published 2012 by John Wiley & Sons, Ltd.